PRAISE FOR TERENCE WRIGHT

From *"The Power of Hallucinogens:"*

The author provides very detail information on both the physical and historical aspects of each hallucinogen. The book gives you the risks and the benefits of each substance as well ... I especially like the personal stories towards the end. I highly recommend this book to anyone who has an interest in hallucinogens and their properties.

— MISSY RICHARDSON

This book provides a comprehensive and informative overview of the history and use of psychedelics. The author does a great job of presenting both the benefits and the potential risks of these substances in a balanced and objective manner.

— STOICPRENEUR

From *"The Power of Dissociatives:"*

I work in law enforcement and see what these drugs (like PCP) do to the people who use them. This book explains both sides, highlighting why someone would feel compelled to take them but what it does to the body ... It was a good read with lots of good info on what hallucinogens do to the person and the society that abuses them. Great book for someone interested in medicine, like a pharmacy student or someone like me who often needs to explain to others what drugs do.

— LYNN COSTA

THE POWER OF HALLUCINOGENS AND DISSOCIATIVES

THE POWER OF HALLUCINOGENS AND DISSOCIATIVES

THE BENEFITS, RISKS, AND MYSTERIES OF 12 POWERFUL PSYCHEDELIC DRUGS (2-IN-1 COLLECTION)

JOURNEY INTO THE PSYCHEDELIC MIND

TERENCE WRIGHT

Book
Bound Studios

This book is dedicated to all those who have experienced the power of hallucinogens and dissociatives, whether for spiritual, therapeutic, or recreational purposes. Your experiences have shaped the understanding and appreciation of these substances, and your journeys have inspired us to explore the depths of our minds and the mysteries of the universe.

Reality is merely an illusion, albeit a very persistent one.

— ALBERT EINSTEIN

CONTENTS

INTRODUCTION

Introducing *"The Power of Hallucinogens and Dissociatives,"* a fascinating and informative 2-in-1 collection from the books *"The Power of Hallucinogens"* and *"The Power of Dissociatives."* This comprehensive guide explores the mind-altering effects of two distinct classes of drugs used for centuries for religious and spiritual purposes and has recently gained popularity recreationally. In this book, you will discover the historical background, chemical properties, medical uses and abuses, risks and side effects, and laws and regulations of these powerful substances. Personal experiences with these drugs are also shared to give readers a deeper understanding of their effects. *"The Power of Hallucinogens and Dissociatives"* is a must-read for anyone interested in learning about the fascinating world of these mind-altering drugs, their impact on mental health, society, and addiction, and how they can be used to enhance our understanding of the human mind.

THE POWER OF HALLUCINOGENS

A GUIDE TO THE HISTORY AND USE OF PSYCHEDELICS, INCLUDING LSD, PSILOCYBIN (MAGIC MUSHROOMS), MESCALINE (PEYOTE), DMT, AND AYAHUASCA

INTRODUCTION

Hallucinogens, also referred to as psychedelics, are a unique class of psychoactive substances utilized in various cultures throughout history for spiritual and medicinal purposes. These powerful compounds alter perceptions, thoughts, and emotions, often leading to profound and transformative experiences.

In this book, we delve into the rich history, effects, and cultural significance of five specific hallucinogens: lysergic acid diethylamide (LSD), psilocybin, mescaline (peyote), N,N-dimethyltryptamine (DMT), and ayahuasca. Each of these substances has unique characteristics and has been used in different ways throughout history. We will explore the scientific and historical aspects of each of these substances, from their discovery and early research to their current uses and potential future applications. We will also look at these substances' cultural and spiritual significance, from their traditional use in indigenous societies to their current use in modern spiritual and therapeutic settings. Finally, we look at a simple guide to taking these psychedelics safely.

We will delve deeper into the effects of these substances on the brain and their potential therapeutic uses, such as for treating addiction, PTSD, and depression. We will also explore their use's risks and side effects and the current legal and policy implications. The book will also feature personal stories and accounts of individuals who have used these substances, providing a unique perspective on their effects and potential benefits.

A vibrant, swirling kaleidoscope of colors and shapes representing the mind-bending effects of hallucinogens.

The book aims to comprehensively understand hallucinogens, shedding light on their complex history, cultural significance, effects, and potential therapeutic uses. In addition, it aims to provide readers with an in-depth understanding of these fascinating substances and the potential they hold for the field of mental health and well-being.

Definition and Background of Hallucinogens

Hallucinogens are a diverse group of psychoactive compounds that can alter an individual's perceptions, thoughts, and emotions. They can be derived from various sources, including plants and animals, and even synthesized in a laboratory. Consequently, the effects of these substances can vary greatly depending on the substance consumed, the dosage, and the individual's state of mind.

Hallucinogens have been used for centuries in various cultures for spiritual and medicinal purposes. The earliest use of hallucinogens dates back to prehistory, with psychoactive substances in ritual and spiritual practices well-documented in many ancient cultures. Traditional medicine has also used these substances for their believed healing properties.

Hallucinogens can be divided into two main categories: classic psychedelics and dissociative psychedelics. Classic psychedelics include lysergic acid diethylamide (LSD), psilocybin, and mescaline (peyote), which enhance sensory experiences and increase emotions. In contrast, dissociative psychedelics such as ketamine disrupt the brain's normal functioning and alter the sense of reality.

The effects of hallucinogens on the brain and how they affect perception thought, and emotion still needs to be fully understood. However, recent research has begun to shed light on the mechanisms by which these substances affect the brain. For example, studies have shown that hallucinogens can increase activity in certain brain regions and disrupt the normal functioning of neural networks, leading to changes in perception, thought, and emotion.

Overall, the term "hallucinogen" encompasses various psychoactive compounds used for centuries in various cultures for spiritual and medicinal purposes. Their effects can vary depending on the substance, dosage, and the individual's state of mind. Despite their historical use, our understanding of these substances' effects and mechanisms of action still needs to be improved. More research is needed to understand their potential therapeutic benefits and risks fully.

Overview of the Five Hallucinogens Discussed in this Book

Hallucinogens, also known as psychedelics, have been used for centuries for spiritual and medicinal purposes. These powerful compounds alter perceptions, thoughts, and emotions, often leading to profound and transformative experiences. In this book, we will explore the history, effects, and cultural significance of five specific hallucinogens: lysergic acid diethylamide (LSD), psilocybin, mescaline (peyote), N,N-dimethyltryptamine (DMT), and ayahuasca. Each of these substances has unique characteristics and has been used in different ways throughout history.

In this chapter, we will delve into the specific properties of these substances. We will begin by exploring LSD, a synthetic compound developed in 1938 and known for its powerful and long-lasting hallucinations. We will then move on to psilocybin, a naturally occurring psychedelic compound found in certain mushrooms, which has been used for centuries by indigenous cultures for spiritual and medicinal purposes and is being studied for its potential therapeutic uses in treating conditions such as depression, anxiety, and PTSD.

We will also examine mescaline, a psychoactive alkaloid found in the peyote cactus. It has been used in Native American spiritual ceremonies for centuries. It is known for its ability to induce vivid and meaningful hallucinations.

Next, we will explore DMT, a naturally-occurring psychedelic substance found in plants and animals. It is typically vaporized or smoked for its powerful but short-lasting effects, which are known to induce intense and mystical experiences.

Lastly, we will look at ayahuasca, a traditional Amazonian brew made from the ayahuasca vine and other plants, known for its strong visual hallucinations and spiritual effects and increasingly being used in therapeutic settings for the treatment of various mental health conditions.

As we will discover throughout this book, each substance has unique properties that make them distinct. Understanding these specific properties can provide a deeper understanding of hallucinogens and their potential benefits for mental health and well-being.

1

LYSERGIC ACID DIETHYLAMIDE (LSD)

Lysergic acid diethylamide (LSD), also known as acid, blotter, dots, and yellow sunshine, is a powerful psychedelic substance that has been the subject of much controversy and debate throughout its history. Developed in 1938 by a Swiss chemist, Albert Hofmann, LSD is a synthetic compound known for its ability to alter perception, thoughts, and feelings in profound ways. The chemical structure of LSD is composed of a ring of four carbon atoms with a nitrogen atom and a diethylamide group attached. It acts on the serotonin receptors in the brain, specifically the 5-HT2A receptor, leading to the alteration of sensory perception, emotions, and thought processes.

In the 1950s and 60s, LSD became popular as a tool for self-discovery and personal growth and played a significant role in the counterculture movement of the time. However, as drug use and abuse increased, concerns about the safety and potential risks of LSD led to its criminalization in the 1970s. Since then, LSD has been classified as a Schedule I controlled substance in the United States, which means it is considered to have a high potential for abuse and no currently accepted medical use.

Despite its criminalization, LSD research has continued. Recent studies have suggested that it may have therapeutic potential for conditions such as cluster headaches, end-of-life anxiety, and alcoholism. This chapter will explore the chemical structure, effects on the brain, history, cultural significance, and therapeutic uses of LSD. It will also provide an in-depth understanding of the specific properties of this substance and its potential therapeutic benefits and risks.

A person with closed eyes sits in a field of vibrant wildflowers, their mind and senses are heightened as they see a kaleidoscope of colors and patterns, they feel the soft petals brush against their skin, the grass rustle beneath them and the gentle breeze blowing on their face, they are lost in an LSD-induced euphoria.

Chemical Structure and Effects on the Brain

Lysergic acid diethylamide (LSD) is a synthetic compound known for its powerful and long-lasting hallucinations. It was first synthesized in 1938 by Albert Hofmann, a Swiss chemist, who is also credited with discovering the psychoactive effects of the compound.

The chemical structure of LSD consists of a ring of four carbon atoms with a nitrogen atom and a diethylamide group attached. The specific structure of LSD is what gives it its unique psychoactive properties. It acts on the serotonin receptors in the brain, specifically the 5-HT2A receptor. This leads to altered sensory perception, emotions, and thought processes.

A scientist in a lab coat, peering through a microscope at a glowing, neon-colored molecule of LSD, symbolizing the future of research and understanding of this powerful psychedelic.

LSD produces profound changes in consciousness, including perception, thought, and emotion. It can cause an individual to experience visual and auditory hallucinations lasting up to 12 hours or more. The effects of LSD can be highly variable and are influenced by factors such as dose, set, and setting. At low doses, the effects of LSD may be subtle. Still, at higher doses, the hallucinations can be vivid and intense.

The effects of LSD on the brain are still not fully understood, but research has begun to shed light on how this substance affects the brain. For example, studies have suggested that LSD affects how the brain processes information and increases communication between different brain regions. Studies have also shown that LSD affects how the brain processes information from the eyes and increases brain plasticity.

In conclusion, LSD is a synthetic compound known for its powerful and long-lasting hallucinations. Its chemical structure and effects on serotonin receptors in the brain make it unique among other psychedelics. Despite its unique properties, the effects of LSD on the brain and its specific mechanisms of action are not fully understood. More research is needed to understand the full effects of LSD and its potential therapeutic benefits.

History and Cultural Significance

Lysergic acid diethylamide (LSD) was synthesized in 1938 by a Swiss chemist, Albert Hofmann while working for the pharmaceutical company Sandoz. At the time, the compound was intended to be used as a circulatory and respiratory stimulant. Still, it was soon discovered to have profound psychedelic properties. Hofmann discovered the compound's psychoactive effects after accidentally absorbing a small amount through his skin while working on the substance.

In the following years, LSD was widely used by scientists, artists, and the general public as a tool for self-discovery and personal growth. It was particularly popular in the 1950s and 60s when it became a central part of the counterculture movement. The counterculture movement embraced LSD to challenge traditional social norms, and the psychedelic experience was seen as a way to expand the mind and gain a new perspective on the world.

However, as drug use and abuse increased, concerns about the safety and potential risks of LSD began to rise. This led to the criminalization of LSD in the 1970s. It has been classified as a Schedule I controlled substance in the United States, which means it is considered to have a high potential for abuse and no currently accepted medical use.

Despite its criminalization, the use of LSD has not disappeared, and it continues to be used today in various contexts. From its historical significance as a powerful tool for self-discovery and personal growth to its cultural significance as a symbol of counterculture, LSD has played a significant role in shaping various aspects of society. Its impact can be seen in areas such as art, literature, music, and the field of psychology, where its use in research and therapy has led to new insights into the workings of the mind.

An old, worn-out book and a pair of round-lensed glasses, sitting on top of a vintage record player, playing psychedelic music from the 60s, representing the historical and cultural significance of LSD in shaping the counterculture and music scene of that era.

It's worth noting that recent studies and clinical trials conducted by various organizations and researchers looking into the therapeutic benefits of psychedelics have had positive results. It is promising in treating various mental health conditions such as depression, PTSD, addiction, and end-of-life anxiety. These renewed interests have led to an ongoing debate about the classification and legal status of psychedelics and their potential benefits for mental health and well-being.

Overall, the history and cultural significance of LSD is a complex and multifaceted subject that spans several decades. From its initial discovery and early uses as a tool for self-discovery and personal growth to its criminalization and continued use today in recreational and therapeutic settings. It's still being debated, and its full potential is not yet understood. Still, the recent studies and research on the therapeutic benefits of psychedelics provide new insights into its use and impact on society.

Risks and Potential Therapeutic Uses

Lysergic acid diethylamide (LSD) is a powerful psychedelic substance, and its effects can be unpredictable. While it can produce profound changes in consciousness that may be beneficial in certain contexts, there are also risks associated with its use.

A person sitting on a hospital bed with a bandaged hand, holding a dropper bottle and a worried expression, with a view of a brain scan on the background, representing the potential therapeutic uses and the risks associated with the use of LSD.

One of the primary risks of LSD is its ability to cause intense and vivid hallucinations that can be disorienting and overwhelming for some individuals. It can also increase the likelihood of experiencing anxiety, paranoia, and psychosis in some people, particularly those predisposed to such conditions. In addition, long-term use of LSD can lead to persistent psychosis, or "flashbacks," which can occur weeks or even months after the last dose. LSD can also cause physiological effects such as increased blood pressure, heart rate, and body temperature, which can be dangerous if not monitored properly.

Despite the negative risks, recent research has also shown potential therapeutic uses of LSD. Studies have suggested that it may be useful in treating conditions such as cluster headaches, end-of-life anxiety, and alcoholism. In addition, research also suggests that it may have a role in treating mental disorders such as depression and anxiety. However, more research is needed to understand the therapeutic benefits and risks of LSD fully. In addition, it's important to note that FDA does not approve LSD as a therapeutic agent. As such, its use for therapeutic purposes is highly restricted.

Finally, LSD is a powerful psychedelic substance with complex effects. While it has the potential to be therapeutic in certain conditions, it also has negative risks and potential for abuse. More research is needed to fully understand the benefits and risks of using LSD as a therapeutic agent. In addition, it should be used under the guidance of a trained medical professional in a controlled and safe environment.

A Guide to Taking LSD

LSD, also known as lysergic acid diethylamide, is a powerful psychedelic drug that can produce intense and profound changes in perception, thought, and emotion. It is available in various forms, including tablets, capsules, and liquids.

The recommended dosage for LSD is typically between 50 and 150 micrograms. However, it is important to remember that individual sensitivity to the drug can vary greatly. Hence, it is best to start with a lower dose and work your way up.

LSD can last anywhere from 6 to 12 hours or more, depending on the dose and the individual. The effects of the drug typically peak around 2-4 hours after ingestion and gradually decrease over the next several hours.

It is recommended to take LSD in a safe and comfortable environment, preferably with people you trust and with whom you feel comfortable. Some people prefer to take the drug outdoors in nature, while others prefer a more controlled setting, such as their own home.

Taking LSD safely involves being prepared for the potential psychological effects of the drug, as well as taking steps to minimize the risk of physical harm. This includes having a sober trip sitter present, staying hydrated, and avoiding dangerous activities while under the influence. It is also important to be aware that LSD can cause flashbacks, so it is best to avoid taking the drug if you have a history of mental health issues or are currently experiencing emotional distress.

It is illegal to consume, possess, and distribute LSD in most countries. Therefore, its possession, use, and distribution are criminal offenses. Therefore, it is important to consider the legal risks before deciding to take the drug.

It is also important to note that this information is for educational purposes only and is not intended as a substitute for professional medical advice, diagnosis, or treatment. It is not safe to take LSD, and one should always consult a healthcare professional before taking any drug.

In conclusion, lysergic acid diethylamide (LSD) is a powerful psychedelic substance with a unique chemical structure and mode of action. Its effects on the serotonin receptors in the brain can cause intense and long-lasting hallucinations that can alter perception, thoughts, and feelings.

Throughout history, LSD has played a significant role in shaping culture, art, literature, and music, as well as in the field of psychology. It was widely used as a tool for self-discovery and personal growth in the 1950s and 60s and played a central role in the counterculture movement of the time. However, as drug use and abuse increased, concerns about the safety and potential risks of LSD led to its criminalization in the 1970s.

Recent research suggests that when used in a controlled setting and under the guidance of trained professionals, LSD has potential therapeutic benefits for treating conditions such as cluster headaches, end-of-life anxiety, and alcoholism. However, it's important to note that more research is needed to understand the therapeutic benefits and risks of LSD fully. In addition, despite its potential therapeutic benefits, it's important to remember that the FDA does not yet approve LSD as a therapeutic agent, which poses risks to its users. As a result, it's important to use caution and fully understand the drug's effects before attempting to use it.

Chapter Summary

- Lysergic acid diethylamide (LSD) is a powerful psychedelic substance that profoundly alters perception, thoughts, and feelings.
- Developed in 1938 by a Swiss chemist, Albert Hofmann, LSD is a synthetic compound that acts on the serotonin receptors in the brain, specifically the 5-HT2A receptor.
- In the 1950s and 60s, LSD became popular as a tool for self-discovery and personal growth and played a significant role in the counterculture movement of the time.
- However, as drug use and abuse increased, concerns about the safety and potential risks of LSD led to its criminalization in the 1970s.
- Despite its criminalization, LSD research has continued. Recent studies have suggested that it may have therapeutic potential for conditions such as cluster headaches, end-of-life anxiety, and alcoholism.
- The effects of LSD can be highly variable and are influenced by factors such as dose, set, and setting. It can cause an

individual to experience visual and auditory hallucinations lasting up to 12 hours or more.

- The effects of LSD on the brain are still not fully understood, but research has begun to shed light on how this substance affects the brain.
- More research is needed to understand the full effects of LSD and its potential therapeutic benefits.

2

PSILOCYBIN

Psilocybin, also known as magic mushrooms, shrooms, and psychedelic mushrooms, is a naturally occurring psychedelic compound found in certain species of mushrooms. The use of these mushrooms has been recorded for centuries in traditional shamanic and spiritual practices by indigenous cultures worldwide. However, it wasn't until the 1950s that psilocybin was first chemically synthesized, and Western scientists discovered its psychoactive properties.

Psilocybin works by acting on the serotonin receptors in the brain, specifically the 5-HT2A receptor, which alters sensory perception, emotions, and thought processes. It produces profound changes in consciousness, including perception, thought, and emotion. The effects of psilocybin can be highly variable and are influenced by factors such as dose, set, and setting.

In recent years, there has been renewed interest in studying the therapeutic potential of psilocybin, particularly for treatment-resistant depression, anxiety, and PTSD. However, despite the promising research, much remains to be understood about the specific mechanisms of action and risks associated with using psilocybin. This chapter will explore the chemical structure, effects on the brain, history, cultural significance, and therapeutic uses of psilocybin.

A person sitting in a forest surrounded by mushrooms, with a serene expression and closed eyes, representing the natural origin and the spiritual and therapeutic potential of Psilocybin.

Chemical Structure and Effects on the Brain

Psilocybin is a naturally occurring psychedelic compound in certain mushrooms, most notably the Psilocybe genus. The chemical structure of psilocybin is composed of an indole ring, a basic structure found in many naturally occurring compounds, and a phosphorylated hydroxy group. This unique combination of elements is what gives psilocybin its psychoactive properties.

Psilocybin acts on the serotonin receptors in the brain, specifically the 5-HT2A receptor, which plays a critical role in regulating mood, emotion, and perception. The activation of the 5-HT2A receptors leads to changes in the neural activity in the brain, resulting in the alteration of sensory perception, emotions, and thought processes.

Psilocybin produces profound changes in consciousness and affective states, including changes in perception, thought, and emotion. As a result, it can cause an individual to experience visual and auditory hallucinations lasting up to 6 hours or more. The effects of psilocybin can be highly variable and are influenced by factors such as dose, set, and setting. At low doses, the effects of psilocybin may be subtle. Still, at higher doses, the hallucinations can be vivid and intense.

A scientist in a lab coat, holding a petri dish with glowing Psilocybin mushrooms, with a background of a brain scan and test tubes, representing the future of research and understanding of the effects of Psilocybin on the brain.

The effects of psilocybin on the brain are not fully understood. Still, research has begun to shed light on the mechanisms by which this substance affects the brain. For example, studies have suggested that psilocybin increases the connectivity between different brain regions, specifically between the default mode network involved in self-referential and reflective thinking and the task-positive network involved in perception, attention, and movement.

History and Cultural Significance

Psilocybin mushrooms have been used for centuries in traditional shamanic and spiritual practices by indigenous cultures worldwide. The use of these mushrooms has been depicted in rock paintings and artifacts, suggesting that their use is a longstanding tradition. The use of psilocybin mushrooms was primarily used for spiritual and medicinal purposes, allowing the shaman or spiritual leader to communicate with the spirit world, access altered states of consciousness, and diagnose and treat illnesses.

An ancient Aztec mushroom stone carving and a modern person taking Psilocybin in a ritual setting, surrounded by candles and incense, representing the historical and cultural significance of Psilocybin use in indigenous cultures and spiritual practices.

It was not until the 1950s that psilocybin was first chemically synthesized by Western scientists, and its psychoactive properties were discovered. This sparked a renewed interest in psilocybin and its potential medical uses among the scientific community. In the following decades, psilocybin research became popular among scientists. It was also used by counterculture movements, who embraced psychedelics as a tool for personal growth, self-discovery, and a way to challenge traditional social norms.

In recent years, there has been renewed interest in the therapeutic potential of psilocybin, particularly for treatment-resistant depression, anxiety, and PTSD. Studies have shown that the use of psilocybin in combination with therapy can lead to significant improvement in symptoms and overall well-being. It's a possible alternative treatment for patients who have not responded to traditional treatments. The recent trend of renewed interest in the study of psychedelics has led to a growing body of research on the therapeutic uses of psychedelics and their safety and efficacy for treating various mental health conditions.

Risks and Potential Therapeutic Uses

A therapist sitting with a patient in a cozy room, holding a Psilocybin capsule and a notebook, with a background of a brain scan and a plant, representing the potential therapeutic uses and the need for responsible use and guidance of Psilocybin in treating mental health issues.

While psilocybin is considered less risky than other psychedelics, it still poses risks to the user. High doses can cause anxiety, paranoia, and other psychological distress. It's also important to note that consuming wild-picked mushrooms can be risky as some mushrooms can be toxic and cause serious harm or even death. Long-term use of psilocybin is also poorly understood and could have a negative impact.

However, research suggests that when used in a controlled setting and under the guidance of trained professionals, psilocybin has a range of therapeutic benefits. For example, studies have shown that psilocybin-assisted therapy can be effective in treating treatment-resistant depression and anxiety, as also PTSD. Furthermore, research suggests that psilocybin may have a therapeutic potential on other mental and physical conditions such as cluster headaches, addiction, and even end-of-life anxiety. Additionally, psilocybin could be used as an aid to psychotherapy, providing an opportunity for personal insight and emotional processing.

While the research is promising, more studies are needed to fully understand the therapeutic benefits and risks of psilocybin and the optimal dosing and administration methods. Despite the ongoing research and promising results, psilocybin is not yet an FDA-approved therapeutic agent and should be used only under the guidance of trained healthcare professionals. Moreover, it is important to be aware of laws and regulations, as psilocybin is a controlled substance in many countries.

A Guide to Taking Psilocybin

Psilocybin is a naturally occurring psychedelic compound found in certain species of mushrooms. It is often used for spiritual or therapeutic purposes. It is known for its ability to create profound changes in the user's perception, thoughts, and emotions. Here is a guide to taking psilocybin safely and effectively.

Psilocybin can be found in several forms, including fresh or dried mushrooms, capsules, or powdered extract. The mushrooms can vary in appearance, but they generally have a brown or tan cap and stem. Capsules and powdered extract are usually sold in small, labeled packages.

The amount of psilocybin needed can vary depending on the individual. Still, a typical starting dose is around 0.5-1.5 grams of dried mushrooms. However, it's important to start with a low dose and work your way up to avoid overwhelming effects.

The effects of psilocybin typically last around 4-6 hours or more, with the experience peak occurring around 2-3 hours after ingestion.

Psilocybin should be taken in a safe, comfortable, and familiar environment. A quiet, dark room or outdoor space can be ideal, as it allows for greater introspection and contemplation. However, it's important to avoid taking psilocybin in public or in an unfamiliar setting, as it can be disorienting and potentially dangerous.

Here are some important tips to keep in mind when taking psilocybin:

- Avoid combining with other substances, such as alcohol or other drugs, as this can increase the risk of negative side effects.
- Make sure you're in a good state of mind before taking psilocybin. Avoid taking it if you're feeling anxious, depressed, or stressed.
- Have a designated "sober" person to monitor you and ensure your safety.
- Be prepared for the possibility of intense emotional experiences or hallucinations.
- Remember that psilocybin can cause confusion and disorientation, so avoid driving or operating heavy machinery while under the influence.

Psilocybin is a powerful substance that can change the user's perception and emotions. To ensure a safe and effective experience, start with a low dose, take it in a comfortable and familiar environment, and have a designated "sober" person around to monitor your safety. With these guidelines in mind, you can experience the benefits of psilocybin while minimizing the risks.

. . .

In conclusion, psilocybin is a naturally occurring psychedelic compound found in certain mushrooms that have been used for centuries in traditional shamanic and spiritual practices by indigenous cultures. Its chemical structure and mode of action on the serotonin receptors in the brain can cause hallucinations and alter perception, thoughts, and feelings.

In recent years, there has been renewed interest in studying the therapeutic potential of psilocybin, particularly for treatment-resistant depression, anxiety, and PTSD. Studies have shown that Psilocybin-assisted therapy can effectively treat these conditions in a controlled setting and under the guidance of trained professionals. Research also suggests that it may be useful in addressing cluster headaches, addiction, and end-of-life anxiety.

While the therapeutic benefits of psilocybin are promising, it's important to remember that more research is needed to understand the therapeutic benefits and risks of psilocybin fully. In addition, the FDA does not approve it as a therapeutic agent, and consuming wild-picked mushrooms can be risky as some mushrooms can be toxic and cause serious harm. Therefore, it's important to use caution and fully understand the drug's effects before attempting to use it.

Overall, psilocybin is a unique substance that has a long history of traditional use and is showing promise in recent research for therapeutic applications. However, it is important to approach its use with care and caution.

Chapter Summary

- Psilocybin is a naturally occurring psychedelic compound found in certain species of mushrooms that have been used for centuries in traditional shamanic and spiritual practices by indigenous cultures worldwide.
- The chemical structure of psilocybin is composed of an indole ring and a phosphorylated hydroxy group, which gives it its psychoactive properties.

- Psilocybin works by acting on the serotonin receptors in the brain, specifically the 5-HT2A receptor, which alters sensory perception, emotions, and thought processes.
- The effects of psilocybin can be highly variable and are influenced by factors such as dose, set, and setting.
- In recent years, there has been renewed interest in studying the therapeutic potential of psilocybin, particularly for treatment-resistant depression, anxiety, and PTSD.
- The use of psilocybin mushrooms has been depicted in rock paintings and artifacts, suggesting that their use is a longstanding tradition. The use of psilocybin mushrooms was primarily used for spiritual and medicinal purposes.
- Psilocybin was first chemically synthesized by Western scientists in the 1950s, and its psychoactive properties were discovered. This sparked renewed interest in its potential medical uses among the scientific community.
- In recent years, there has been renewed interest in studying the therapeutic potential of psilocybin, particularly for treatment-resistant depression, anxiety, and PTSD. However, much remains to be understood about the specific mechanisms of action and risks associated with using psilocybin.

3

MESCALINE (PEYOTE)

Mescaline, also known as peyote or "mesc," is a psychoactive alkaloid found in the peyote cactus. It has been used for centuries in the traditional spiritual practices of indigenous people in Mexico and the American Southwest. Mescaline is known for its ability to induce vivid and meaningful hallucinations. In addition, western scientists and artists have used it to explore the nature of consciousness.

In this chapter, we will explore the chemical structure and effects of mescaline on the brain, its history and cultural significance, and its potential therapeutic uses and risks. Understanding the specific properties of this substance and its potential benefits can provide a deeper understanding of the broader topic of psychedelics.

Chemical Structure and Effects on the Brain

Mescaline, also known as 3,4,5-trimethoxyphenethylamine (or 3,4,5-TMPEA for short), is a naturally occurring psychoactive alkaloid found in the peyote cactus (Lophophora williamsii) and other cacti species. It is known for its ability to induce vivid and meaningful hallucinations, as well as its ability to change the way an individual perceives the world.

A person sitting cross-legged in front of a cactus with a dreamy expression, surrounded by a swirling aura of vibrant colors, representing the natural origin and the spiritual and therapeutic potential of Mescaline.

Mescaline's chemical structure consists of a phenethylamine core, a type of organic compound found in many naturally occurring substances, with a methoxy group attached to the nitrogen R3 and an isopropyl group attached to the nitrogen R2. The specific arrangement of atoms in the molecule gives it its unique psychoactive properties.

When consumed, mescaline acts on the serotonin receptors in the brain, specifically the 5-HT2A receptor, responsible for regulating mood, perception, and cognitive processes. This leads to altered sensory perception, emotions, and thought processes. Depending on the individual, the effects can vary depending on the dosage and can last anywhere from 8 to 12 hours.

At low doses, the effects of mescaline can be subtle. They may include changes in perception of colors and shapes and a heightened sense of well-being and euphoria. At higher doses, the hallucinations can be vivid and intense. They may include changes in perception of time and a sense of interconnectedness with the world. Additionally, mescaline may have potential therapeutic uses for treating certain mental health conditions like depression, anxiety, and PTSD. Still, it's important to note that more research is needed to understand the therapeutic benefits and risks of the substance fully.

A researcher in a lab coat, holding a test tube containing a bright orange liquid with a background of a brain scan and scientific equipment, representing the future of research and understanding of the effects of mescaline on the brain.

History and Cultural Significance

The use of mescaline and the peyote cactus holds deep cultural and spiritual significance for indigenous people, who have used it for centuries in traditional ceremonies and rituals. The peyote cactus, also known as the "sacred cactus," is central to the belief systems and religious practices of many indigenous tribes. In addition, its use is often seen as a way to connect with the spiritual world and gain a deeper understanding of one's place.

In addition to its traditional use, Western scientists and artists have also widely studied mescaline. In the early 20th century, mescaline was used in experimental psychology and psychiatry as a tool to explore the nature of consciousness and perception. Researchers believe the psychedelic experiences induced by mescaline provide insight into the workings of the human mind. They used it in research on schizophrenia and other mental disorders.

Ancient rituals and modern cultural significance of mescaline in a trippy, surreal landscape of shifting colors and shapes.

The counterculture movement of the 1950s and 60s also embraced mescaline and other psychedelics to challenge traditional social norms and expand the mind. During this time, mescaline was used by artists and writers as a tool for self-expression and creativity. However, as with other psychedelics, mescaline's increasing use and abuse led to concerns about its safety and potential risks.

In conclusion, mescaline is a psychoactive alkaloid known for its ability to induce vivid and meaningful hallucinations. Its use has deep cultural and spiritual significance for indigenous people and has played a significant role in Western science and counterculture history.

Understanding mescaline's cultural significance and history are crucial to fully understanding its effects on human consciousness and potential therapeutic uses.

Risks and Potential Therapeutic Uses

While mescaline is considered relatively safe, with few negative side effects when used in traditional, controlled settings, its effects can vary depending on the individual, the dosage, and the setting in which it is consumed. For example, high doses can lead to anxiety, paranoia, and psychological distress. Furthermore, as with any psychedelic substance, there is a risk of having a "bad trip," which can lead to lasting negative psychological effects.

A psychedelic cactus in the desert, its vibrant colors swirling as a person under its influence navigates the blurred lines between therapeutic treatment and dangerous consequences.

Despite the lack of scientific research, traditional use of peyote cactus and anecdotal evidence suggest potential therapeutic benefits of mescaline. Some studies have suggested that mescaline-assisted therapy may be useful in treating addiction, depression, anxiety, and PTSD. Research also suggests that it may have a role in treating cluster headaches and end-of-life anxiety. More controlled, rigorous research is needed to understand mescaline's therapeutic benefits and risks fully, and it is not approved as a therapeutic agent by the FDA. It's important to note that consuming wild-harvested peyote may be risky as it could be misidentified or contaminated.

A Guide to Taking Mescaline

Mescaline is a psychedelic substance found in certain cacti, most notably the peyote cactus. It can also be synthesized in a laboratory. The substance is typically consumed by ingesting the cactus but can also be consumed in powder or capsule form.

When taking mescaline, it is important to know that the effects can vary depending on the dose consumed. A moderate dose can range from 200-500mg, while a heavy dose can be over 1000mg. The effects of mescaline typically last for 8-12 hours.

It is important to take mescaline in a safe and comfortable environment, ideally with people you trust. A safe and comfortable setting will help to minimize the risk of a bad trip or other negative reactions.

It is also important to remember that mescaline can produce intense hallucinations and alter your perception of reality, so it is important to have a clear state of mind and be in a state of well-being before consuming it.

As with any psychedelic substance, it is important to start with a low dose and work your way up, as the effects can be unpredictable. It is also essential to be aware that the use of mescaline is illegal in many places. Hence, it is important to be aware of the laws in your area.

It is also important to be aware that mescaline can interact with other medications you might be taking, so it is essential to consult with a doctor or a medical professional before consuming it.

In summary, mescaline is a powerful psychedelic substance that can be consumed in various forms. Before taking it, it is important to be aware of the dosage, the duration of the effects, the environment, and the risks. Taking it in a safe and comfortable place with people you trust and a clear state of mind is important. Please consult with a doctor or a medical professional before consuming it.

In conclusion, mescaline is a psychoactive alkaloid found in the peyote cactus that has a long history of use in traditional spiritual practices by indigenous cultures. It's known for its ability to induce vivid and meaningful hallucinations. In addition, its effects on the brain, specifically on serotonin receptors, lead to altered sensory perception, emotions, and thought processes. Despite its historical use, there is limited scientific research on the therapeutic benefits of mescaline. However, studies have suggested its potential therapeutic benefits for treating addiction, depression, and anxiety. Still, more research is needed to understand mescaline's therapeutic benefits and risks fully. Additionally, it's important to note that mescaline is not an FDA-approved therapeutic agent and should be used under the guidance of trained professionals. However, mescaline's cultural and historical significance cannot be denied, adding a unique dimension to understanding its effects and potential future use.

Chapter Summary

- Mescaline, also known as peyote, is a psychoactive alkaloid found in the peyote cactus.
- It has been used for centuries in the traditional spiritual practices of indigenous people in Mexico and the American Southwest.
- Mescaline is known for its ability to induce vivid and meaningful hallucinations.
- Mescaline's chemical structure is a phenethylamine core, with a methoxy group attached to the nitrogen R3 and an isopropyl group attached to the nitrogen R2.

- Mescaline acts on serotonin receptors in the brain, specifically the 5-HT2A receptor, which is responsible for regulating mood, perception, and cognitive processes.
- The effects of mescaline can vary depending on the dosage and can last anywhere from 8 to 12 hours.
- The use of mescaline and the peyote cactus holds deep cultural and spiritual significance for indigenous people.
- Western scientists and artists have also widely studied mescaline in the early 20th century as a tool to explore the nature of consciousness and perception. The counterculture movement of the 1950s and 60s also embraced mescaline and other psychedelics.

4

N,N-DIMETHYLTRYPTAMINE (DMT)

N,N-Dimethyltryptamine, also known as DMT, Dimitri, The Spirit Molecule, Businessman's Trip, Dimenstyline, and N,N-DMT, is a naturally-occurring psychedelic substance in various plants and animals. This powerful substance is known for its intense and mystical effects, often described as profound and life-changing. Due to its unique properties, DMT has gained much attention in neuroscience and psychotherapy.

In this chapter, we will take a closer look at the chemical structure of DMT and how it interacts with the brain. We will also explore DMT's historical and cultural significance, including its use in traditional spiritual practices. Furthermore, we will examine the potential therapeutic benefits of DMT, including its use in treating addiction, PTSD, and other mental health conditions.

However, it is important to note that DMT is not without risks. The use of DMT can lead to adverse effects, including hallucinations, paranoia, and even psychosis. Therefore, it is essential to approach the use of DMT with caution and to be fully informed about the potential risks and benefits associated with its use.

A journey through the neon-lit, otherworldly landscape of the DMT experience, as a traveler navigates the intense and mind-bending terrain of this powerful psychedelic.

As we delve deeper into this topic, we will better understand the unique properties of DMT and its place in the broader world of psychedelics. Despite the potential risks, the therapeutic benefits of DMT warrant further research and investigation. Understanding the chemical structure, effects on the brain, historical and cultural significance, therapeutic benefits, and risks associated with DMT can help us make more informed decisions about its use.

Chemical Structure and Effects on the Brain

N,N-Dimethyltryptamine (DMT) is a naturally-occurring psychedelic substance found in many plants and animals, including certain species of trees, shrubs, and mushrooms. Its chemical structure is composed of a tryptamine molecule with a dimethyl group attached to the nitrogen atoms, which gives it its unique psychoactive properties. DMT acts on the serotonin receptors in the brain, specifically the 5-HT2A receptor, leading to the alteration of sensory perception, emotions, and thought processes.

A scientist in a lab coat, peering through a microscope at the molecular structure of DMT as it illuminates in vibrant colors, unlocking the secrets of this powerful psychedelic and its potential therapeutic uses.

DMT is typically vaporized or smoked for its powerful and short-lasting effects. Its onset is rapid, and the effects can be intense and profound. The experience can include vivid visual hallucinations, profound changes in thought and emotion, and a sense of euphoria. In addition to its powerful hallucinations, DMT is also known to induce mystical and spiritual experiences. Some users have reported experiencing a sense of transcendence and connection to a higher power. The duration of the effects of DMT is relatively short, usually lasting around 30-60 minutes.

DMT is considered one of the most powerful psychedelics. However, its effects can be highly variable and are influenced by dose, set, and setting factors. While DMT is considered relatively safe, it can also pose some risks to the user. For example, high doses can cause anxiety, paranoia, and other psychological distress. In addition, some users may experience psychologically distressing effects known as "bad trips." Therefore, it's important to use DMT under the guidance of trained professionals in a controlled setting.

History and Cultural Significance

DMT has a long history of use in traditional spiritual practices by indigenous cultures in South America. It is consumed in the form of a brew called ayahuasca. However, its use in the Western world began in the 1950s. It was popularized by writers such as Terence McKenna and Rick Strassman, who wrote extensively about their experiences with the substance.

A DMT shaman in traditional headdress, surrounded by burning sage and ancient tribal artifacts, guiding a group of seekers on a journey through the otherworldly realms of DMT culture.

DMT has also been used in various spiritual and religious practices worldwide, with different cultures having unique ways of consuming it. The effects of DMT are considered to be profound. They can include intense and vivid hallucinations and feelings of transcendence. It is known to produce mystical experiences, and some users report encountering entities or otherworldly beings. DMT has been associated with various spiritual practices, including meditation, shamanism, and ceremonial rituals.

DMT has also played a role in the counterculture movement of the 1960s and 1970s. It was considered a powerful tool for self-discovery and personal growth. Many artists and musicians used it to access new forms of creativity. DMT has also been studied by scientists and researchers who have sought to understand its effects on the brain and consciousness.

Criminalized in the 1970s, DMT is considered a Schedule I controlled substance in the United States, with no accepted medical use and a high potential for abuse. However, recent studies have shown the potential therapeutic benefits of DMT, particularly in treating addiction and PTSD, but research is still ongoing.

Risks and Potential Therapeutic Uses

DMT, while considered to be relatively safe in terms of physical risks, its effects on the mind can be intense and potentially distressing for some individuals. High doses or long-term use can lead to persistent psychosis or "flashbacks," which can occur weeks or even months after the last dose. It's also important to note that DMT is a Schedule I controlled substance. Therefore, it is illegal to manufacture, possess, or use DMT in most countries.

However, in recent years, research has begun to explore the potential therapeutic benefits of DMT. For example, studies have suggested that DMT-assisted therapy may be useful in treating addiction, PTSD, and depression. Additionally, DMT has been used in spiritual and therapeutic settings to achieve personal growth and self-discovery. However, research on the therapeutic benefits of DMT is still in its early stages, and more research is needed to understand its potential therapeutic uses and risks fully.

Lastly, DMT is a naturally-occurring psychedelic substance with a long history of use in traditional spiritual practices. Its chemical structure, effects on the brain, historical and cultural significance, potential therapeutic uses, and risks make it a fascinating subject to study. While more research is needed to understand the therapeutic benefits of DMT fully, and it is not legal and not an approved therapeutic agent by FDA, it has the potential for further exploration in the mental health and well-being field.

A person in the middle of a DMT trip, the room is filled with vibrant, pulsing lights and fractal patterns as they journey through the intense and mind-bending realms of the DMT experience.

A Guide to Taking DMT

DMT is a powerful psychedelic substance that can produce intense and profound experiences. It can be found in various forms, including powder, crystals, and liquid, and can be ingested, vaporized, or injected.

The appropriate dosage of DMT varies depending on the form, method of consumption, and individual tolerance. A typical oral dosage is between 20-40 milligrams, while vaporization typically requires a lower dose of around 10-20 milligrams. Therefore, it's important to start with a low dosage and gradually increase it to avoid overwhelming experiences or negative reactions.

A DMT experience can vary but typically lasts around 30-60 minutes when vaporized and 1-2 hours when consumed orally.

When taking DMT, it's important to have a comfortable and safe environment. It's recommended to have a trusted sitter present and avoid taking DMT in public or where you feel unsafe.

The safety of DMT use largely depends on the method of consumption, dosage, and individual tolerance. It can have serious short-term and long-term effects on mental and physical health when consumed in high doses or when mixed with other substances. Therefore, before consumption, it is important to research the substance and its positive and negative effects. It is also important to remember that DMT is a Schedule I controlled substance in the United States and many other countries, and possession and use are illegal.

It is important to remember that the effects of psychedelics can vary greatly from person to person, and it is important to approach them with caution and care. As always, it's best to consult a healthcare professional before consuming any substance.

In conclusion, DMT is a naturally-occurring psychedelic substance found in plants and animals used for centuries in traditional spiritual practices. Its chemical structure consists of a tryptamine molecule with a dimethyl group at the nitrogen positions, which allows it to interact with serotonin receptors in the brain and produce intense and mystical experiences.

Despite its long history of use in traditional settings, scientific research on the effects of DMT on the brain is still limited. However, studies have suggested that DMT may have therapeutic potential in treating certain conditions such as addiction, PTSD, and depression. However, more research is needed to understand DMT's potential benefits and risks fully. In addition, it's important to note that DMT is illegal in many countries and not approved as a therapeutic agent by the FDA.

Overall, the study of DMT provides insight into the complex and unique effects of psychedelics on the human mind and consciousness. Further research on the potential therapeutic uses of DMT could lead to new treatments for mental health conditions and a deeper understanding of the nature of consciousness.

Chapter Summary

- N,N-Dimethyltryptamine (DMT) is a naturally-occurring psychedelic substance found in many plants and animals, including certain species of trees, shrubs, and mushrooms.
- Its chemical structure is composed of a tryptamine molecule with a dimethyl group attached to the nitrogen atoms, which gives it its unique psychoactive properties.
- DMT acts on the serotonin receptors in the brain, specifically the 5-HT2A receptor, leading to the alteration of sensory perception, emotions, and thought processes.
- DMT is typically vaporized or smoked for its powerful and short-lasting effects, including vivid visual hallucinations, profound changes in thought and emotion, and a sense of euphoria.
- DMT has a long history of use in traditional spiritual practices by indigenous cultures in South America and was popularized by writers in the Western world in the 1950s.
- DMT has played a role in the counterculture movement of the 1960s and 1970s and has been studied by scientists and researchers to understand its effects on the brain and consciousness.
- DMT is considered a Schedule I controlled substance in the United States, with no accepted medical use and a high potential for abuse. Still, recent studies have shown the potential therapeutic benefits of DMT, particularly in treating addiction and PTSD.

- DMT, while considered to be relatively safe in terms of physical risks, its effects on the mind can be intense and potentially distressing for some individuals. They should be used under the guidance of trained professionals in a controlled setting.

5

AYAHUASCA

In this chapter, we will delve into the fascinating world of ayahuasca—a traditional Amazonian brew used for centuries by indigenous people for spiritual, medicinal, and ceremonial purposes. The brew, made from the ayahuasca vine (Banisteriopsis caapi) and other plants, is known for its powerful psychedelic effects and is believed to possess a wide range of therapeutic benefits.

We will explore the history and cultural significance of ayahuasca, its chemical composition, and its effects on the brain. Additionally, we will examine the potential therapeutic uses of ayahuasca and the risks associated with its use. The chapter also will include recent research developments in the field and the legal status of ayahuasca. By the end of this chapter, you will have a comprehensive understanding of this fascinating brew and its place in modern society.

Chemical Structure and Effects on the Brain

Ayahuasca, a traditional Amazonian brew, combines the ayahuasca vine (Banisteriopsis caapi) with other plants, particularly the leaves of the chacruna plant (Psychotria Viridis), which contain the powerful psychedelic compound DMT. This brew has been used for centuries by indigenous people for spiritual and medicinal purposes.

A vibrant jungle scene with a lone figure sitting cross-legged in front of a boiling pot of Ayahuasca, surrounded by the glowing eyes of curious animals peeking out from the foliage.

The chemical composition of ayahuasca is particularly noteworthy, as it contains two key ingredients that interact in the body to produce its psychoactive effects. The first active ingredient is DMT, a naturally-occurring compound known to produce strong visual hallucinations and a sense of euphoria and spiritual transcendence. The second active ingredient is MAOIs (Monoamine oxidase inhibitors), which are found in the ayahuasca vine.

When consumed, DMT and MAOIs interact with the brain in unique ways. DMT works by binding to serotonin receptors in the brain, which releases other neurotransmitters that alter the perception of reality. Conversely, the MAOIs inhibit the breakdown of neurotransmitters, allowing DMT to remain active in the body for longer periods.

Combining these two ingredients creates a powerful synergy that produces various psychological and perceptual effects associated with ayahuasca. The brew is known to elicit a broad range of emotions, which can range from a profound sense of euphoria and well-being to feelings of intense fear and anxiety. The exact nature of the effects of ayahuasca can depend on the dose consumed, the context in which it is consumed, and the individual's state of mind and prior experiences.

A scientist in a lab coat stands in front of a microscope, studying a sample of the Ayahuasca plant while on the screen behind him, abstract chemical structures and neural pathways light up, representing the inner workings of this mysterious plant and its effects on the human brain.

Due to its strong psychoactive effects, ayahuasca has been the subject of much scientific research. It is considered to have potential therapeutic applications in psychiatry and psychology. Despite this, the use of ayahuasca is controversial, and its legal status varies from country to country.

History and Cultural Significance

Ayahuasca, a traditional Amazonian brew, has a rich and complex history that spans centuries of use by indigenous people in the region.

The brew is considered to be a sacred medicine. It has been used for various purposes, including spiritual, medicinal, and ceremonial practices. Indigenous people believe that ayahuasca can connect with the spirit world, achieve spiritual insight and understanding, and access hidden knowledge and wisdom.

An ancient temple hidden deep in the Amazon jungle, where a shaman is performing a traditional Ayahuasca ceremony for a group of indigenous people, surrounded by intricate carvings and hieroglyphs depicting the rich history of the spiritual practice.

The use of ayahuasca has been passed down through generations and has deep cultural roots in indigenous communities. It is often used in traditional ceremonies that are steeped in ritual and symbolism and which are considered to be powerful and transformative experiences. The use of ayahuasca is closely tied to indigenous people's spiritual beliefs and practices, and it is seen as an essential part of their traditional way of life.

In the 20th century, the use of ayahuasca began to spread outside of the Amazon region, particularly in the Western world. As more people became interested in the brew and its potential benefits, it began to be used in various new contexts. For example, in the Western world, ayahuasca has been used in spiritual and therapeutic settings as a tool for self-discovery and personal growth and to treat various mental health conditions.

The cultural significance of ayahuasca is reflected in the growing number of people who travel to the Amazon to participate in traditional ceremonies to gain a deeper understanding of the indigenous culture and its practices. In addition, the popularity of ayahuasca tourism has increased awareness of the brew and its potential benefits and raised several ethical questions about commercializing traditional practices.

The scientific community is increasingly recognizing the cultural significance of ayahuasca. Studies on the neuropsychopharmacology of ayahuasca have revealed that it can be used as a tool for healing and self-discovery and plays an important role in preserving the cultural heritage of the Amazonian people. In recent years, ayahuasca has also gained the attention of scientists, anthropologists, and other researchers studying its use and the benefits it can provide.

Risks and Potential Therapeutic Uses

While considered a powerful and transformative tool, ayahuasca carries certain risks and potential dangers. For example, consuming ayahuasca can lead to psychological distress in some individuals, including anxiety, paranoia, and intense hallucinations. It's also important to note that, while physical risks are considered to be low when consumed under the guidance of a trained facilitator, long-term use or high doses of ayahuasca can lead to persistent psychosis, or "flashbacks," which can occur for weeks or even months after the last dose.

Furthermore, certain medications may interact negatively with ayahuasca, such as antidepressants and blood thinners, and they should not be combined with those medications. Moreover, it's not recommended for people with a history of serious mental illnesses and other medical conditions, so it's important to consult a healthcare professional before consumption.

A person lies on a mat in a dimly lit room, closed eyes, and a look of peaceful surrender on their face, while swirling shapes and colors dance in their mind's eye, as they embark on a journey of self-discovery and introspection under the guidance of an Ayahuasca trip.

Despite its risks, ayahuasca has also been studied for its potential therapeutic benefits. Studies have suggested that ayahuasca-assisted therapy may be useful in treating conditions such as addiction, PTSD, and depression. Research has also suggested that it may have benefits for treating symptoms of anxiety, depression, and PTSD in individuals with treatment-resistant conditions. Additionally, ayahuasca is being researched for its potential to help alleviate symptoms of PTSD, anxiety, and depression.

However, research on the therapeutic benefits of ayahuasca is still relatively limited, and more studies are needed to understand its potential benefits and risks fully. In many countries, ayahuasca is considered a Schedule I drug, and it's illegal to possess, consume or sell it. However, in some other countries, it's regulated and used in therapeutic contexts. Additionally, it's not approved by the FDA as a therapeutic agent. Therefore, it's important to be aware of the legal status of ayahuasca in one's country and only uses it under a trained professional's guidance.

A Guide to Taking Ayahuasca

Ayahuasca is a powerful psychoactive brew traditionally used by indigenous people in the Amazon region for spiritual and healing purposes. It is typically made by boiling the ayahuasca vine (Banisteriopsis caapi) and the leaves of the chacruna plant (Psychotria Viridis) together.

Different forms of ayahuasca can vary in color, consistency, and strength, but typically the liquid is dark brown and thick. The taste is often described as bitter and unpleasant.

The recommended dosage of ayahuasca varies depending on the individual and the specific brew used. It's important to start with a small dose and gradually increase it to find your comfort level. It's also important to be guided by an experienced facilitator.

The effects of ayahuasca can last anywhere from 4 to 8 hours. It can cause intense hallucinations, changes in perception, and powerful emotional experiences.

Ayahuasca ceremonies are typically held in a traditional setting, such as a maloca (a large round thatched-roof building) in the Amazon jungle. However, it can be taken in other settings, such as a retreat center or a private ceremony.

Safety is a major concern when taking ayahuasca. Therefore, working with a qualified and experienced facilitator or shaman who can provide guidance and support throughout the experience is important. Additionally, it's important to avoid certain medications, such as antidepressants and blood thinners, and to inform the facilitator of any pre-existing medical conditions. It's also important to be in a safe and comfortable environment and to have a trusted sitter or guide who can assist you during the experience.

It's also important to note that ayahuasca is illegal in some countries, and it's important to check the laws in your area before attempting to take it.

. . .

In conclusion, ayahuasca is a traditional Amazonian brew used for centuries by indigenous people for spiritual, medicinal, and ceremonial purposes. Its unique chemical composition, containing DMT and MAOIs, allows for powerful psychedelic effects on the brain. Research has suggested potential therapeutic uses for addiction, PTSD, and depression. Still, more research is needed to understand the benefits and risks associated with its use fully.

The cultural significance of ayahuasca reflects the growing number of people who travel to the Amazon to participate in traditional ceremonies, heal themselves, and understand the indigenous people's spiritual and ceremonial practices. However, it's also important to note that ayahuasca carries certain risks, including psychological distress and potential negative interactions with certain medications, and should only be consumed under the guidance of a trained professional.

Ayahuasca is a complex substance with many potential benefits and risks. This chapter aims to provide a comprehensive overview of the history, cultural significance, chemical composition, effects, therapeutic uses, and risks associated with ayahuasca to help readers understand this fascinating brew. However, it's important to remember that even though ayahuasca has a lot of potential benefits, it's only recommended for some, and one should be aware of its legal status and the potential risks before consuming it.

Chapter Summary

- Ayahuasca is a traditional Amazonian brew made from the ayahuasca vine and other plants. It is known for its powerful psychedelic effects and is believed to possess therapeutic benefits.
- The brew's chemical composition includes the active ingredient DMT, a naturally-occurring compound known to produce strong visual hallucinations and a sense of euphoria, and MAOIs (Monoamine oxidase inhibitors) found in the ayahuasca vine.

- When consumed, DMT and MAOIs interact with the brain uniquely, releasing other neurotransmitters that alter the perception of reality.
- The effects of ayahuasca can range from euphoria to intense fear and anxiety, depending on the dose, context, and individual's state of mind.
- Ayahuasca has potential therapeutic applications in psychiatry and psychology. Still, its use is controversial, and its legal status varies by country.
- Ayahuasca has a rich history of use by indigenous people for spiritual, medicinal, and ceremonial purposes. It is considered a sacred medicine with deep cultural roots.
- In the 20th century, ayahuasca began to spread outside of the Amazon region, particularly in the Western world, where it is used in spiritual and therapeutic contexts.

6

PERSONAL STORIES

This chapter focuses on personal stories of individuals who have used psychedelics, specifically lysergic acid diethylamide (LSD), psilocybin, mescaline (peyote), N,N-dimethyltryptamine (DMT), and ayahuasca for various purposes.

These stories offer a unique and personal perspective on the effects of these substances on the mind and how they can be used for personal growth, therapy, spiritual exploration, and other purposes. Through the stories, you will be able to understand the diversity of experiences and the potential benefits and risks of using psychedelics.

Ultimately, this chapter serves as a window into the subjective experience of these individuals and the profound ways in which psychedelics can impact one's life and consciousness.

Stories About LSD

John's Story of Using LSD for Personal Growth

John had always been deeply invested in personal development, constantly seeking new and innovative ways to improve himself and gain a greater understanding of the world around him. As he delved deeper into this pursuit, he became increasingly curious about the effects of psychedelics on the mind and the potential benefits they could offer those seeking self-improvement.

A kaleidoscope of colors swirls around John's mind as he reflects on his journey of self-discovery through the use of LSD, each fragment of light represents a newfound realization and personal growth, illuminating the path towards a brighter future.

After much research and contemplation, John finally decided to take the plunge and try LSD for the first time under the guidance and supervision of a trained professional in a safe and controlled setting. He was aware of the potential risks and hazards but also felt confident in his readiness and ability to handle the experience.

As soon as the LSD took effect, John knew he was in for something special. The world around him seemed to take on a whole new dimension, and he felt his mind opening up to new and previously unimaginable possibilities. He described the experience as profound and transformative, allowing him to gain insight into his thought patterns and behaviors in ways he could never have imagined.

Through the journey, John felt like he was able to overcome

personal barriers and achieve a greater sense of self-awareness. The insight and understanding he gained from experience allowed him to see his life and himself in a new light. They helped him to navigate the path of personal growth with more grace and ease. He felt that the LSD experience had been a crucial step in his self-discovery journey, and he has been grateful for the opportunity to explore the inner workings of his mind in such a profound way.

Sarah's Experience With LSD-Assisted Therapy

Sarah had been dealing with the debilitating effects of anxiety and depression for many years. Despite her best efforts, she could not find lasting relief through traditional therapy and medication. She had exhausted all the options and was feeling hopeless and at a loss as to what to do next. That's when she learned about LSD-assisted therapy as a potential alternative.

At first, Sarah was skeptical and hesitant. She had heard the stories about the dangers and risks associated with LSD, and she wasn't sure if this was a path she wanted to take. But the more she researched, the more she learned about the promising results of this form of therapy. So finally, she saw it as a possible last resort worth exploring.

Sarah was under the guidance of a trained therapist, who helped her prepare for the therapy sessions and explained the process at every step. As a result, she felt comfortable in a safe and controlled environment and could relax and trust the process.

As soon as the LSD took effect, Sarah felt her mind start to open up. She found that the LSD helped her to see her thoughts and emotions from a new perspective and that it helped her to process and understand them differently. In addition, the therapy sessions unlocked a door inside her mind, allowing her to access and confront parts of herself that she had been trying to avoid or suppress.

Sarah reported feeling more relaxed and at ease after the therapy sessions. She found that her depression and anxiety symptoms had greatly improved and that she had gained a greater understanding and insight into her struggles. She was grateful for the opportunity to try LSD-assisted therapy and saw it as a life-changing experience that helped her regain control of her mental and emotional well-being.

Stories About Psilocybin

Emily stands in a serene forest, surrounded by mushrooms, she reaches out to touch one, as memories of her traumatic past flash before her eyes, but instead of feeling overwhelmed, she feels a sense of healing and peace wash over her as the psilocybin works its magic, helping her overcome her PTSD.

Michael's Journey Using Psilocybin for Spiritual Exploration

Michael had always been fascinated by spirituality and the search for meaning and purpose in life. However, he had struggled to find a traditional practice or belief system that truly resonated with him and felt like it fit. He had tried many different approaches over the years. Still, nothing seemed to give him the depth of understanding and connection he was searching for.

In his quest for something deeper, Michael first learned about the potential of psychedelics, specifically psilocybin, as a means of exploring spirituality. He had heard stories of others having profound spiritual experiences. At the same time, under the influence of these substances, he was intrigued by the possibility that it could be a way for him to access the deeper understanding and connection he was searching for.

After much research and contemplation, Michael decided to take the plunge and try psilocybin in a formal setting with a trained guide. He wanted to approach the experience with the utmost respect and reverence, and he felt this was the best way to do so.

As soon as he consumed the psilocybin, Michael entered a profound and mystical experience. He felt his mind opening up to new and previously unimaginable possibilities and a deep sense of connection with a deeper sense of self and the universe. He described the experience as transformative, allowing him to gain insight into his thought patterns and behaviors in ways he could never have imagined. In addition, he felt he could tap into a sense of interconnectedness and oneness that he had never experienced before.

Through the journey, Michael found that the experience helped him better understand himself and his place in the world. He felt like he had been given a new perspective on his existence and that he had gained a deeper understanding of the spiritual nature of reality. In addition, he had found a new way of connecting to a spiritual dimension, which helped improve his life. He felt that the psilocybin experience had been a crucial step in his spiritual journey, and he's been grateful for the opportunity to explore the inner workings of his mind and soul in such a profound way.

Emily's Account of Using Psilocybin to Overcome PTSD

Emily had been living with the debilitating effects of Post-Traumatic Stress Disorder (PTSD) for several years. Despite her best efforts, she could not find lasting relief through traditional therapy and medication. Instead, she has tried different types of therapy, such as cognitive-behavioral and talk therapy. Still, she found that they did not help her to address the root cause of her PTSD. She had exhausted all the options and was feeling hopeless and at a loss as to what to do next. That's when she learned about psilocybin-assisted therapy as a potential alternative treatment for PTSD.

At first, Emily was skeptical and hesitant. She had heard the stories about the dangers and risks associated with psychedelics, and she wasn't sure if this was a path she wanted to take. But the more she researched, the more she learned about the promising results that had been achieved with this form of therapy, particularly for those with PTSD. So finally, she saw it as a possible last resort worth exploring.

Emily was under the guidance of a trained therapist, who helped her prepare for the therapy sessions and explained the process at every step. As a result, she felt comfortable in a safe and controlled environment and could relax and trust the process.

As soon as the psilocybin took effect, Emily felt her mind start to open up. She found that psilocybin helped her to process and understand her traumatic memories differently. It allowed her to gain insight and understanding into the root causes of her PTSD and helped her to let go of the negative thoughts, emotions, and memories associated with her traumatic experiences. In addition, the therapy sessions unlocked a door inside her mind, allowing her to access and confront parts of herself that she had been trying to avoid or suppress.

Emily reported feeling more relaxed and at ease after the therapy sessions. She found that her PTSD symptoms had greatly improved and that she had gained a greater understanding and insight into her struggles. She was grateful for the opportunity to try psilocybin-assisted therapy and saw it as a life-changing experience that had helped her overcome the negative impact of her traumatic experiences on her life and regain control of her emotional well-being.

Stories About Peyote

Scott's Account of Using Peyote in a Native American Ceremony

Scott had always been drawn to the rich culture and spiritual practices of Native American tribes. He had spent many years studying and learning about their ways of life. He had always felt a deep connection to the natural world. He believed that the traditional practices of these ancient cultures held many answers to his questions about the meaning and purpose of existence.

Scott sits cross-legged in a circle of tribal elders, the desert landscape stretching out behind him, in his hand he holds a small cactus button.

As part of his ongoing quest to learn more about Native American spirituality, Scott decided to participate in a peyote ceremony led by a traditional Native American medicine man. Peyote is a small, spineless cactus with the psychoactive alkaloid mescaline, which has been used for centuries by indigenous people of North America for spiritual and ceremonial purposes.

The ceremony took place throughout the night, and other participants and traditional songs and prayers accompanied Scott. He was honored to be allowed to participate in such a sacred and meaningful ceremony. He approached the experience with humility and respect.

As soon as he consumed the peyote, Scott felt himself entering into a realm of profound and spiritual experience. He felt his mind opening up to new and previously unimaginable possibilities and a deep connection with the natural world and himself. He described the experience as transformative, allowing him to gain insight into his thought patterns and behaviors in ways he could never have imagined. In addition, he felt he could tap into a sense of interconnectedness and oneness that he had never experienced before.

Through the journey, Scott felt like he had been given a new perspective on his existence and that he had gained a deeper understanding of the spiritual nature of reality. He felt like he had been able to connect with the natural world in a way that was deeper and more meaningful than he had ever thought possible. He felt that the peyote ceremony had been a crucial step in his spiritual journey. He felt grateful for the opportunity to participate in such a sacred and meaningful ceremony. In addition, he was honored to have been able to connect with the teachings of the ancient indigenous cultures of North America.

Rachel's Experience With Peyote in a Therapeutic Setting

Rachel had been struggling with addiction for many years. Despite her best efforts, she could not find lasting relief through traditional therapy and medication. Instead, she has tried different types of therapy, such as cognitive-behavioral and talk therapy. Still, she found that they did not help her to address the root cause of her addiction. She had exhausted all the options and was feeling hopeless and at a loss as to what to do next. That's when she learned about peyote-assisted therapy as a potential alternative treatment for addiction.

At first, Rachel was skeptical and hesitant. She had heard the stories about the dangers and risks associated with psychedelics, and she wasn't sure if this was a path she wanted to take. But the more she researched, the more she learned about the promising results that had been achieved with this form of therapy, particularly for those with addiction. She began to see it as a possible last resort worth exploring.

Rachel was under the guidance of a trained therapist, who helped her prepare for the therapy sessions and explained the process at every step. As a result, she felt comfortable in a safe and controlled environment and could relax and trust the process.

As soon as the peyote took effect, Rachel felt her mind start to open up. She found that the peyote helped her to understand the root causes of her addiction and the underlying emotional and psychological factors that had led to it. It allowed her to gain insight into her addiction and the patterns of thoughts and behaviors that had kept her trapped in its cycle. The therapy sessions unlocked a door inside her mind, allowing her to access and confront parts of herself that she had been trying to avoid or suppress.

Rachel reported feeling more in control of her life after the therapy sessions. She found that her addiction symptoms had greatly improved and that she had gained a greater understanding and insight into her struggles. She was grateful for the opportunity to try peyote-assisted therapy and saw it as a life-changing experience that helped her overcome her addiction and regain control of her life.

Stories About DMT

David's Ecounter With an "Entity" During a DMT Trip

David had always been deeply interested in the realm of psychedelics and the potential benefits they could offer for exploring the nature of consciousness and the self. He had read extensively about the various substances and their effects. He had always been particularly intrigued by DMT's powerful and intense effects.

Samantha sits in a clearing surrounded by a brilliant array of wildflowers as the sun sets on the horizon, she closes her eyes taking in a deep breath of the fresh clean air and begins to inhale the smoke of DMT.

After much research and contemplation, David finally decided to take the plunge and try DMT for himself, under the guidance and supervision of a trained professional in a safe and controlled setting. He was aware of the potential risks and hazards but also felt confident in his readiness and ability to handle the experience.

As soon as the DMT took effect, David knew he was in for something special. He described the experience as intense and otherworldly. He felt as though he had been transported to a completely different dimension. As the trip progressed, he felt like he had encountered an "entity" or intelligent being communicating with him in a language he couldn't understand. David felt like the entity had shared deep insights and knowledge with him. He felt like the entity was trying to convey important information. Still, the language barrier prevented him from fully understanding.

David felt that the experience had been deeply meaningful and transformative and had changed his perspective on life and consciousness in ways he could never have imagined. He felt that the DMT experience had helped him gain a new understanding of reality and the nature of existence. It had opened his mind to new and unimaginable possibilities. He felt that the experience was a glimpse into the hidden depths of reality and consciousness. It left him with a sense of wonder and awe about the world and our place.

Samantha's Use of DMT to Connect With Nature

Samantha had always felt a profound connection to nature. She had always found solace and inspiration in the beauty of the natural world. From an early age, she had been drawn to the outdoors, spending much of her free time hiking, camping, and exploring the wilderness.

As she got older, Samantha began to feel the desire to explore this connection to nature more profoundly. As a result, she became curious about the potential benefits of psychedelics. In particular, she had heard about DMT's powerful and transformative effects and felt that it might deepen her connection to nature in ways she had never experienced before.

After much research and contemplation, Samantha decided to take the plunge and try DMT for herself in the most natural setting, surrounded by the beauty of nature that could be a place for her to immerse into the DMT experience fully. She found a place where she felt comfortable and could relax and trust the process.

As soon as the DMT took effect, Samantha felt her connection to nature intensifying and felt like she had been transported to a different realm. The beauty of the natural world became even more vivid and intense, and she felt as though she had connected with the natural world in a new and deeper way. She described the experience as profound and felt that it helped her to understand the interconnectedness of all things and the oneness of nature.

Samantha felt that the experience had strengthened her connection to nature and her sense of self. She felt that the DMT experience had helped her understand reality and the nature of existence and opened her mind to new and unimaginable possibilities. She felt that the experience was a glimpse into the hidden depths of reality and consciousness. It left her with a sense of wonder and awe about the world and our place. She felt that the experience had helped her to deepen her appreciation and understanding of the natural world and to reconnect with the beauty of the natural world in a way that was deeper, more meaningful, and more profound than she had ever thought possible.

Stories About Ayahuasca

Anthony embraces the healing power of the Ayahuasca ceremony surrounded by the jungle's lush greenery and the comforting flames of a small fire, the Ayahuasca vine's vines symbolize his connection to the earth and the ancestors who guide him on his journey to healing.

Anthony's Journey to Healing Through the Ayahuasca Ceremony

. . .

Anthony had been struggling with addiction for many years. Despite his best efforts, he could not find lasting relief through traditional therapy and medication. He had tried different types of therapy, such as cognitive-behavioral and talk therapy, as well as support groups and 12-step programs. Still, he found that they did not help him address his addiction's root cause. He had exhausted all of the available options and was feeling hopeless and at a loss as to what to do next.

That's when he heard about ayahuasca, a powerful psychedelic plant medicine used for centuries by indigenous people of the Amazon for spiritual and healing purposes. He learned about its potential for helping individuals overcome addiction and other psychological issues and was deeply intrigued by the idea of using it to heal.

Anthony decided to participate in an ayahuasca ceremony led by a traditional shaman who had decades of experience working with the medicine. He knew that this would be a powerful and potentially challenging experience. Still, he was determined to do whatever it took to overcome his addiction and find lasting healing.

The ceremony took place in a traditional setting, and other participants and traditional songs and prayers accompanied Anthony. As soon as he drank the ayahuasca brew, he felt himself entering a realm of intense and emotional experience. He felt like the ayahuasca helped him to confront and overcome the root causes of his addiction by allowing him to understand the underlying emotional and psychological factors that had led to it.

Anthony felt that the ceremony helped him to gain a greater sense of self-awareness and control. He felt that he had a better understanding of himself and his addiction and that he had been able to let go of the negative thoughts, emotions, and memories that had kept him trapped in the cycle of addiction. He felt he had been given the tools and insights to take control of his healing and move forward positively.

During the ceremony, he felt like he had been able to connect with his innermost self and gain a deeper understanding of the origins of his addiction. He felt like he could see things from a different perspective, allowing him to see the patterns and behaviors that had led him to addiction and kept him trapped. He felt that the ayahuasca had given him the courage and strength to confront these patterns and behaviors head-on and start healing.

After the ceremony, Anthony felt a sense of renewed hope and optimism. He felt that he had been given a second chance and that he had been given the tools and insights he needed to start building a better life for himself. He felt like he had been given a new perspective on his addiction and that he had gained a deeper understanding of himself and the world around him. He was grateful for the opportunity to participate in the ayahuasca ceremony. He felt it had been a life-changing experience that helped him overcome his addiction and regain control of his life.

Gina's Account of Ayahuasca-Assisted Therapy for Depression

Gina had been struggling with depression for many years. Despite her best efforts, she could not find lasting relief through traditional therapy and medication. Instead, she tried different types of therapy, such as cognitive-behavioral therapy and talk therapy. Still, she found that they did not help her to address the root cause of her depression. She had exhausted all the options and was feeling hopeless and at a loss as to what to do next.

That's when she heard about ayahuasca-assisted therapy as a potential alternative treatment for depression. She learned about how ayahuasca had helped other people understand and process their feelings of depression in a new way and was deeply intrigued by the idea of using it as a means of healing.

Gina decided to participate in ayahuasca-assisted therapy led by a trained therapist, who helped her prepare for the therapy sessions and explained the process at every step. As a result, she felt comfortable in a safe and controlled environment and could relax and trust the process.

As soon as the ayahuasca took effect, Gina felt her mind start to open up. She found that the ayahuasca helped her to understand and process her feelings of depression in a new way by allowing her to gain insight and understanding into the underlying emotional and psychological factors that had led to it. It helped her to understand her depression more deeply and to see it from a different perspective.

Gina reported feeling more positive and hopeful after the therapy sessions. She found that her depression symptoms had greatly improved and that she had gained a greater understanding and insight into her struggles. She was grateful for the opportunity to try ayahuasca-assisted therapy and saw it as a life-changing experience that had helped her to overcome her depression and take back control of her life.

The therapy sessions also helped her to gain a greater understanding of the interconnectedness of all things. The ayahuasca had given her a deeper understanding of the mind-body connection, how emotions and thoughts are connected, and how they affect her overall well-being. This understanding helped her to have a more holistic approach to her healing journey. As a result, she felt she had more tools and insights to address her depression from different angles.

In conclusion, the personal stories in this chapter provide insight into the different effects of the drugs discussed in the book. They add to understanding these substances' potential therapeutic uses and cultural significance. Each story is unique and reflects the individual's personal experience and perception. Still, together they paint a broader picture of the effects and implications of these substances. It's important to note that these substances should be used with caution and should always be conducted under the guidance of a trained professional in a safe and controlled setting.

AFTERWORD

The book delves into the complex and often-misunderstood topic of hallucinogens, focusing on five specific substances: LSD, psilocybin, mescaline (peyote), DMT, and ayahuasca. Each substance is unique and has distinct properties, brain effects, and history.

Starting with LSD, also known as lysergic acid diethylamide, the book examines the chemical structure and how it interacts with the brain to produce vivid hallucinations and profound changes in perception. The book also covers the history of LSD, including its initial discovery in the 1940s, its widespread use in the 1950s and 1960s, and its eventual criminalization in the 1970s.

Next, the book delves into psilocybin, a chemical found in certain species of mushrooms. Psilocybin is structurally similar to serotonin, a neurotransmitter in the brain. It is believed to play a role in regulating mood, perception, and other cognitive processes. The book also covers the history of psilocybin, including its use in traditional indigenous cultures and its recent resurgence in research as a potential therapeutic tool.

Mescaline (peyote) is a naturally occurring psychoactive compound in the peyote cactus. The book covers its chemical structure and effects on the brain and its traditional use in indigenous cultures of Mexico and the southwestern United States.

DMT (N,N-dimethyltryptamine) is a powerful psychedelic drug structurally related to serotonin. The book describes the effects of DMT on the brain and consciousness, its history and cultural significance, and its traditional use in Amazonian shamanic practices.

Lastly, ayahuasca is a brew traditionally used by indigenous people of the Amazon basin. It is made from the ayahuasca vine and chakruna leaves. The book covers the history and cultural significance of ayahuasca, its use in traditional shamanic practices, and its chemical composition and effects on the brain.

In addition to providing information on these substances' physical and historical aspects, and a guide to taking them safely, the book also includes personal accounts of individuals who have used these substances for personal growth, spiritual exploration, and therapeutic purposes. Through these accounts, the book aims to provide a nuanced and well-rounded understanding of these substances' potential risks and benefits. Ultimately, this book aims to foster a deeper understanding and appreciation of these powerful and complex substances and their potential role in human consciousness and well-being.

Overview of Current Research and Areas for Future Study

Recent years have seen an increase in research on the therapeutic uses of hallucinogens. A growing body of evidence suggests that these substances may have potential benefits for treating various mental health conditions. Studies have been conducted using psychedelics such as LSD, psilocybin, and ayahuasca to treat addiction, post-traumatic stress disorder (PTSD), depression, and anxiety. These studies have generally been small-scale, but they have produced promising results. As a result, they have generated significant interest in further research.

Despite this progress, there is still much that remains unknown about the therapeutic uses of hallucinogens. To fully understand the benefits and risks of these substances, large-scale, well-designed clinical trials are needed. These trials should include long-term follow-up to determine treatment effects' durability and active comparison, groups. Additionally, research is needed to understand the optimal dosing, frequency, and duration of treatment.

One key area that requires further research is the mechanisms of action of these substances. While we know that psychedelics interact with the serotonin system in the brain, we do not fully understand how these interactions lead to the changes in perception and consciousness associated with the psychedelic experience. Additional research is needed to identify the specific neural pathways and receptors involved in psychedelics' therapeutic effects.

Furthermore, more research must be conducted on these substances' cultural and spiritual significance. They have been used in traditional and indigenous cultures for centuries, often as part of religious or spiritual practices. However, while these substances may have an important role in these cultures, there is a limited scientific understanding of the mechanisms behind these cultural and spiritual uses, so future research could investigate the cultural and spiritual aspects of these substances by studying their use in traditional or indigenous cultures.

While research on the therapeutic uses of hallucinogens is progressing, much is still to be learned. Additional research is needed to confirm the findings of initial studies, to understand the optimal dosing and administration of these substances, and to fully understand the mechanisms of action and potential risks associated with their use. With continued research, we can gain a deeper understanding of these powerful and complex substances and how they can be used to improve human well-being.

Implications for Society and Policy

The potential therapeutic benefits of hallucinogens have significant implications for society and policy. As discussed in the book, these substances have the potential to be used to treat a variety of mental health conditions, such as addiction, PTSD, and depression. However, due to their current Schedule I controlled substances classification, it isn't easy to research their therapeutic potential. This can make it challenging to fully gather the necessary data to understand their benefits and risks and develop effective and safe treatments. Additionally, this classification can criminalize individuals who use or possess these substances for personal growth or therapeutic reasons.

Given the current state of research, it's important to consider a more nuanced and evidence-based approach to regulating these substances. This could include reclassifying certain substances, such as psilocybin, as Schedule IV controlled substances, which have a lower potential for abuse and are considered to have a currently accepted medical use. This would allow for more research to be conducted and protect individuals who use these substances for therapeutic purposes.

Furthermore, it is important to recognize that this substance could have different levels of risk and benefits. Therefore, different regulations can be applied. Not all Hallucinogens are equal in terms of risk and benefits. Therefore, reclassifying and regulating some substances differently from others might be appropriate.

This book has provided a comprehensive overview of hallucinogens, including their chemical structure and effects on the brain, history and cultural significance, risks, potential therapeutic uses, and a guide to taking them safely. It has also highlighted the importance of further research and a more nuanced approach to regulating these substances. As research continues to evolve, it's crucial to consider the implications of these findings for society and policy. This includes the need to reclassify certain substances, adjust regulations to allow for further research, and ensure that individuals who use these substances for therapeutic purposes are protected.

THE POWER OF DISSOCIATIVES

LEARN ABOUT KETAMINE, PCP, DXM, NITROUS OXIDE, MXE, 3-MEO-PCP, AND SALVIA DIVINORUM AND THEIR IMPACT ON MENTAL HEALTH, SOCIETY, AND ADDICTION

INTRODUCTION

Dissociative drugs, also known as **hallucinogens**, are a group of mind-altering substances that produce changes in perception, consciousness, and behavior. They have been used for centuries for religious and spiritual purposes, but in recent times, they have become popular recreationally. Dissociative drugs have both positive and negative effects on individuals and society, making it crucial to understand their impact.

This book provides a comprehensive overview of common dissociative drugs, including ketamine, PCP, DXM, nitrous oxide, MXE, 3-MeO-PCP, and salvia divinorum. You will learn about their historical background, chemical properties, medical uses and abuses, risks and side effects, and laws and regulations. Personal experiences with dissociative drugs are also presented to give readers a deeper understanding of the effects these substances can have. Furthermore, the book explores the relationship between dissociative drugs and mental health, society, and addiction, providing a well-rounded examination of the topic.

The target audience for this book includes anyone interested in learning about dissociative drugs, healthcare professionals, law enforcement, policymakers, and those seeking treatment for dissociative drug addiction. This book aims to educate readers about the effects of dissociative drugs and provide information that individuals can use to make informed decisions about their use. But first, let's tackle some FAQs to set the context.

What are dissociative drugs?

Dissociative drugs are a class of drugs that produce a feeling of detachment from one's environment and self. They can cause hallucinations, changes in perception, and a sense of detachment from reality.

What are the effects of dissociative drugs?

The effects of dissociative drugs vary depending on the type of drug and the dose taken. Common effects include altered perception of time and space, visual and auditory hallucinations, out-of-body experiences, and feelings of detachment from reality.

What are the risks associated with dissociative drugs?

Dissociative drugs can cause serious physical and psychological health risks, including increased heart rate, blood pressure, body temperature, nausea, vomiting, confusion, and paranoia. In addition, long-term use can lead to addiction, depression, and other mental health issues.

Are there any medical uses for dissociative drugs?

Dissociative drugs have been used in medical settings to treat pain, anxiety, and other psychological disorders. However, they are highly regulated and should only be used under the supervision of a medical professional.

How can I prevent the misuse of dissociative drugs?

The best way to prevent the misuse of dissociative drugs is to educate yourself about the risks and dangers associated with them. If you or someone you know is using dissociative drugs, seek help from a medical professional.

In conclusion, dissociative drugs significantly impact individuals, society, and the world. Therefore, it is crucial to understand the effects and potential dangers. This book provides a comprehensive and engaging examination of this topic, making it an essential resource for anyone interested in learning about dissociative drugs.

1

KETAMINE

An unexpected journey down the Rabbit Hole as you experience the surreal and dreamlike effects of ketamine.

Ketamine, also known as special K, kit kat, cat valium, or simply K, is a powerful anesthetic that has been used medically for decades. Recently, it has gained notoriety for its use as a recreational drug. This chapter aims to provide a comprehensive overview of ketamine, covering its history, chemical properties, medical uses and abuses, risks and side effects, and laws and regulations. By presenting a well-rounded picture of this drug, we aim to give readers the knowledge they need to make informed decisions about ketamine use. Whether you are a curious reader, a medical professional, or someone considering using ketamine, this chapter has something for you. So sit back, relax, and join us as we explore the fascinating world of ketamine.

Historical Background

Ketamine was first synthesized in 1962 by the American pharmaceutical company Parke-Davis. It was developed as a safe, fast-acting anesthetic for **surgical procedures**, particularly when traditional anesthetics were unsuitable. Its unique properties, including its ability to produce a dissociative state in patients, quickly made it popular in medical circles.

In the 1970s and 1980s, ketamine was widely used in veterinary medicine. However, by the late 1980s, it had become a popular recreational drug. This led to increased regulations, restrictions on its use, and concerns about its potential for abuse. Despite this, ketamine continued to be used for medical purposes, particularly in emergency and trauma situations where its rapid onset and short duration of action were valuable.

In recent years, ketamine has seen a resurgence in popularity as a treatment for depression, anxiety, and other mental health conditions. This has led to renewed interest in its potential therapeutic benefits, as well as its potential risks and side effects. With a rich and complex history, ketamine remains one of the most intriguing substances in medicine and drug use.

Chemical Properties and Effects

Ketamine is a synthetic drug belonging to the arylcyclohexylamine class of drugs. It is a powerful anesthetic that acts on the central nervous system and can produce a dissociative state in users. When used medically, it is administered intravenously, intramuscularly, or orally.

The effects of ketamine can vary greatly depending on the dose and the method of administration. At low doses, it can produce mild, *trance-like* effects. In contrast, higher doses can cause hallucinations, dream-like states, and a feeling of detachment from the body and the environment. In addition, the dissociative state produced by ketamine can lead to a loss of consciousness. This property makes it a useful anesthetic in medical procedures.

In addition to its anesthetic properties, ketamine has been shown to have pain-relieving and anti-inflammatory effects, making it useful in managing chronic pain conditions. It is also used as an antidepressant and a treatment for anxiety and other mental health conditions due to its ability to reduce symptoms of depression and rapidly improve mood.

The precise mechanisms by which ketamine produces its effects are not yet fully understood. Still, it is thought to act on specific receptors in the brain that are involved in regulating mood and consciousness. Further research is needed to understand the chemical properties and effects of ketamine fully. Still, its unique combination of therapeutic and recreational effects makes it a substance of ongoing interest and study.

Medical Uses and Abuses

Ketamine is a widely used anesthetic in medical settings. It is prized for its fast onset, short duration of action, and low risk of causing respiratory depression. It is often used in emergency and trauma situations, as well as in children, due to its safety profile. In addition to its anesthetic properties, ketamine has also been used to manage chronic pain conditions and depression, anxiety, and other mental health conditions.

Despite its therapeutic uses, ketamine has also been abused for its hallucinogenic and dissociative effects. In addition, it has been used recreationally for its ability to produce intense and often unpredictable experiences. Its popularity as a party drug has led to concerns about its potential for abuse and addiction.

The abuse of ketamine can lead to various negative health effects, including hallucinations, confusion, paranoia, and impaired motor function. In addition, high doses can cause temporary paralysis, unconsciousness, and respiratory depression, leading to serious health consequences and even death.

To reduce the risk of abuse and negative health effects, ketamine is carefully regulated in many countries, and its use is restricted to medical settings. Despite this, it remains a widely used and abused substance, and efforts are ongoing better to understand its therapeutic benefits and potential for abuse.

Risks and Side Effects

Like all drugs, ketamine risks side effects and adverse reactions. The most common side effects of ketamine use include confusion, agitation, hallucinations, and impaired motor function. In addition, high doses can cause temporary paralysis, unconsciousness, and respiratory depression, leading to serious health consequences and even death.

Long-term use of ketamine can lead to various negative health effects, including bladder dysfunction, liver damage, and cognitive impairment. In addition, evidence suggests that repeated exposure to ketamine can lead to tolerance and dependence and persistent changes in brain function and behavior.

In addition to the physical risks associated with ketamine use, it can also have psychological effects, including anxiety, paranoia, and depression. These effects can be especially pronounced in individuals with a history of mental health conditions or who are particularly sensitive to the effects of drugs.

It is important to be aware of the risks and side effects associated with ketamine use and to use the drug only under the supervision of a medical professional. In addition, individuals considering using ketamine for recreational purposes should be aware of the risks involved and should take steps to minimize these risks and protect their health.

Laws and Regulations

Ketamine is a **controlled substance in many countries**, and its use is restricted to medical settings. The same laws and regulations surrounding ketamine vary by country and jurisdiction. Still, it is generally classified as a Schedule III controlled substance, meaning it is illegal to manufacture, distribute, or possess the drug for non-medical purposes.

In the United States, ketamine is regulated by the Drug Enforcement Administration (DEA) under the Controlled Substances Act. This classification makes it illegal to produce, distribute, or possess ketamine for non-medical purposes. Those caught doing so can face severe legal penalties, including fines and imprisonment.

In other countries, the laws and regulations surrounding ketamine may be even more restrictive. For example, the drug may be classified as a Schedule I or Schedule II controlled substance, with even greater penalties for those caught possessing or distributing the drug illegally.

Despite its strict legal status, ketamine remains a widely used and abused substance, and efforts are ongoing better to understand its therapeutic benefits and potential for abuse. To minimize the risk of negative health effects and legal consequences, it is important to be aware of the laws and regulations surrounding ketamine and to use the drug only under the supervision of a medical professional.

Experience Reports

Welcome to the experience reports section! Here, you can read firsthand accounts from people who have taken ketamine and experienced its effects. Their stories provide valuable insight into the wide range of experiences that can occur when taking this substance – both positive and negative. In addition, these reports will help you understand what to expect when taking ketamine.

Report 1: Ketamine Brings Celebratory Joy

I had heard about Ketamine before, but I was never brave enough to try it until I attended a party recently. I was interested to see what the experience would be like and was curious to find out. I figured why not try it and see what happens.

So, I took the plunge and consumed some Ketamine. At first I felt a little anxious and apprehensive, but soon the worries melted away and I found myself feeling a sense of joy and freedom. I felt my body move to the music in a way that I had never felt before. Every beat seemed to have its own unique rhythm, and I was one with the music and my body was euphorically dancing to its own beat. I felt a sense of connection with everyone around me, as if we were all part of a single energy. My face was stuck in a permanent smile, and I just felt so in tune with the present moment.

The experience was honestly amazing and unlike anything I had ever felt before. I felt so light and free, as if I was floating on a cloud. It was a truly uplifting and intense experience, and I am so glad that I had the courage to try something new. I would definitely recommend Ketamine to anyone who is looking for an unforgettable experience.

Report 2: A Negative Ketamine Experience

A few weeks ago, I decided to try out the drug ketamine. Everyone had told me it was a beautiful, enlightening experience, and I was interested to see for myself. However, when I took it, I was met with a completely different reality. Instead of feeling empowered and enlightened, I was left feeling confused, disoriented and muted. My ability to communicate and make sense of my surroundings had been taken away from me. I was unable to enjoy the experience, and instead, was filled with an intense sense of paranoia and fear. My body felt heavy and I felt like I had no control over myself, making the experience a very negative one for me. I was disappointed that my experience was so different from what I had hoped for, and it left me feeling overwhelmed.

Report 3: A Moment of Intense Creativity Through Ketamine

I had been struggling with a creative block for months, feeling like I was stuck in a rut and unable to move forward with any of the projects I had begun. But after trying ketamine, everything changed. I suddenly had a clarity of thought I hadn't experienced in months -- a clarity that allowed me to see my creative potential in a new light.

The first art piece I created after this experience was unlike any I had done before. I felt as though I had been freed from the chains of my own creativity and was now able to express myself in a way that was true and genuine. I felt inspired and motivated to keep creating, and soon found myself in a flow state I had never experienced before.

I soon realized that my newfound clarity wasn't just a temporary burst of creativity, but rather a permanent shift in my perception. I was now more aware of my own strengths and weaknesses, and had a newfound appreciation for the creative process. I was able to approach each project with a newfound confidence and enthusiasm, knowing that I could create something unique and powerful.

This newfound clarity stayed with me for months, allowing me to explore my creative potential and become a better artist. I was now able to create work that I was truly proud of, and I was constantly inspired and motivated to keep creating. I had found a peace and contentment within myself that I had never experienced before, and I was able to use this newfound knowledge to continue to develop my creative abilities.

My experience with ketamine was life-changing, and has allowed me to explore my creative potential and inner peace in ways that I never could have imagined before. I am forever grateful for this experience, and will continue to use the knowledge I have gained to help me reach new heights in my creative endeavors.

In conclusion, ketamine is a powerful anesthetic with a rich and complex history, ranging from its use in medical procedures to its notoriety as a recreational drug. Its chemical properties and effects on the central nervous system make it a unique substance of interest and study. However, it is important to be aware of the risks and side effects associated with its use, including confusion, agitation, hallucinations, and potentially serious health consequences. Despite its restrictions and regulations, ketamine remains a widely used and abused substance. Further research is needed to understand its therapeutic benefits and potential for abuse fully. Whether you are a curious reader, a medical professional, or considering using ketamine, it is essential to have a well-rounded understanding of this fascinating substance.

Chapter Summary

- Ketamine is a powerful anesthetic used medically for decades, but it is also abused as a recreational drug.
- It was first synthesized in 1962 by Parke-Davis as a safe, fast-acting anesthetic. It gained popularity in medical circles due to its unique properties and ability to produce a dissociative state in patients.
- In the late 1980s, ketamine became a popular recreational drug leading to increased regulations and restrictions on its use.
- Recently, ketamine has seen a resurgence in popularity as a treatment for depression, anxiety, and other mental health conditions.

- It is a synthetic drug belonging to the arylcyclohexylamine class. Its effects vary based on the dose and method of administration, ranging from mild trance-like effects to hallucinations and a feeling of detachment.
- The precise mechanisms of how ketamine produces its effects are not yet fully understood, but it is thought to act on specific receptors in the brain involved in regulating mood and consciousness.
- Despite its therapeutic uses, ketamine is also abused for its hallucinogenic and dissociative effects. Its abuse can lead to negative health effects such as hallucinations, confusion, paranoia, impaired motor function, and even death.
- Its use is restricted to medical settings and carefully regulated. Still, it remains widely used and abused, and there is an ongoing effort to better understand its benefits and potential for abuse.

2

PCP (PHENCYCLIDINE)

The air was thick and heavy as a PCP-fueled haze blanketed the city, a surreal landscape of neon-tinged buildings and alleys. In the distance, a faint hum of bass and synthetic beats could be heard, setting the tone for a night of psychedelic exploration.

Phencyclidine, also known as PCP, angel dust, hog, or love boat, is a powerful and controversial substance with a rich history that spans decades. It is a dissociative anesthetic that was first developed in the 1950s as an intravenous anesthetic for surgical procedures but was eventually **abandoned** due to its unpredictable and sometimes dangerous side effects. Despite this, PCP has found its way into the recreational drug market, where it continues to be used for its mind-altering effects. In this chapter, we will explore the historical background of PCP, its chemical properties and effects, its medical uses and abuses, the risks and side effects associated with its use, the laws and regulations surrounding it, and finally, experience reports from those who have used it. Whether you are a curious reader or a student of drug culture, this chapter offers a comprehensive and engaging look at one of the most enigmatic substances of our time.

Historical Background

The history of Phencyclidine (PCP) dates back to the 1950s when it was first synthesized as an intravenous anesthetic for surgical procedures. Developed by Parke-Davis, a pharmaceutical company, PCP was marketed under the trade name Sernyl and was used as a general anesthetic until 1965. However, society discontinued its widespread use due to reports of unpredictable and dangerous side effects, including hallucinations, delusions, and agitation.

Despite its abandonment as a medical anesthetic, PCP found its way into the recreational drug market in the 1960s and 1970s. It gained a reputation as a potent hallucinogen. Its popularity grew; by the 1980s, PCP was one of the most commonly abused drugs in the United States. This led to increased media attention and a heightened public concern, leading to a crackdown on the production and sale of PCP.

Despite this, PCP remains a significant problem in the world of illegal drugs, and its use remains controversial. While it has been banned in most countries, PCP continues to be manufactured and sold on the black market, often under different names. The historical background of PCP is a story of a potent drug that has been celebrated and reviled. It serves as a cautionary tale of the dangers of recreational drug use.

Chemical Properties and Effects

Phencyclidine (PCP) is a dissociative anesthetic that disrupts communication between the brain and the body. It acts as a selective antagonist at the N-methyl-D-aspartate (NMDA) receptor, responsible for pain perception, learning, and memory.

Chemically, PCP is a white crystalline powder soluble in water and is often sprinkled on marijuana or tobacco. Individuals can take it orally or intravenously. The effects of PCP vary depending on the dose and the method of administration. Still, they can last anywhere from 4 to 24 hours.

At low doses, PCP can cause euphoria, empathy, and detachment from reality. Higher doses can cause symptoms such as hallucinations, delusions, agitation, and paranoia. PCP is also known to cause profound alterations in perception, including changes in body image and the distorted experience of time and space.

In addition to its mind-altering effects, PCP has several physiological effects, including increased blood pressure and heart rate, dry mouth, and decreased body temperature. Unfortunately, it can also cause muscular rigidity and seizures, which can be dangerous and even life-threatening in some cases.

In conclusion, the chemical properties and effects of PCP are complex and far-reaching, and they can profoundly impact the mind and body. Furthermore, the unpredictable nature of PCP and its potential for dangerous side effects highlight the importance of avoiding its use.

Medical Uses and Abuses

Phencyclidine (PCP) was first developed in the 1950s as an intravenous anesthetic for surgical procedures. Still, it was eventually **abandoned** due to its unpredictable and dangerous side effects. Despite its abandonment as a medical anesthetic, PCP found its way into the recreational drug market, which continues to be used for its mind-altering effects.

The use of PCP as a recreational drug has been associated with several negative outcomes, including hallucinations, delusions, and paranoia, as well as physical symptoms such as muscle rigidity, seizures, and an increased risk of overdose. Additionally, PCP is highly addictive, and long-term use can lead to permanent cognitive impairment and behavioral problems.

Despite its well-known risks and dangers, PCP continues to be used and abused by individuals seeking its powerful mind-altering effects. In addition, it is often used with other drugs, such as marijuana or alcohol, which can increase its potential for harm.

In conclusion, the medical uses and abuses of PCP are a testament to the dangers of recreational drug use. While PCP was once developed as a medical anesthetic, its unpredictable and dangerous side effects have led to its abandonment in that field. Its continued use as a recreational drug highlights the importance of avoiding drugs with unknown and potentially harmful effects.

Risks and Side Effects

The use of Phencyclidine (PCP) is associated with several serious risks and side effects, both physical and psychological. These can include:

- **Hallucinations and delusions:** PCP is known to cause profound alterations in perception, including hallucinations, delusions, and paranoia.
- **Agitation and violence:** PCP can cause aggression, anger, and violence, and it has been associated with several violent crimes.
- **Seizures and muscle rigidity:** PCP can cause muscle rigidity, convulsions, and seizures, which can be dangerous and even life-threatening in some cases.

- **Increased heart rate and blood pressure:** PCP can cause a rapid increase in heart rate and blood pressure, which can be particularly dangerous for individuals with heart conditions.
- **Addiction:** PCP is known to be highly addictive, and long-term use can lead to permanent cognitive impairment and behavioral problems.
- **Cognitive impairment:** PCP use can lead to memory loss, confusion, and other cognitive problems that can persist even after the drug has left the body.
- **Psychological distress:** PCP use can cause feelings of anxiety, depression, and other forms of psychological distress.

In conclusion, the risks and side effects of PCP are serious and far-reaching, and they highlight the importance of avoiding its use. Whether you are considering using PCP recreationally or have a history of drug abuse, it is important to understand the potential consequences of its use and to seek professional help if necessary.

Laws and Regulations

Phencyclidine (PCP) is a controlled substance in many countries, including the United States, where it is classified as a Schedule II drug under the Controlled Substances Act. This classification indicates that PCP has a high potential for abuse and limited medical use and is subject to strict controls and regulations.

In the United States, the possession, sale, and distribution of PCP are illegal, and individuals caught with the drug can face serious criminal charges, including imprisonment and fines. In addition, the manufacture and sale of PCP are also regulated, and individuals who engage in these activities can face additional legal consequences.

The laws and regulations surrounding PCP are intended to discourage its use and to minimize the harm associated with its abuse. By making PCP illegal and subject to criminal sanctions, authorities aim to reduce the drug's availability and encourage individuals to seek help if they have a problem with its use.

In conclusion, the laws and regulations surrounding PCP are an important component of efforts to reduce the harm associated with its use. By making PCP illegal and subject to criminal sanctions, authorities aim to discourage its use and minimize the harm associated with its abuse. Suppose you have questions or concerns about PCP or its regulation. In that case, it is important to seek the guidance of a knowledgeable professional.

Experience Reports

Welcome to the experience reports section! Here, you can read first-hand accounts from people who have taken PCP (Phencyclidine) and experienced its effects. Their stories provide valuable insight into the wide range of experiences that can occur when taking this substance – both positive and negative. In addition, these reports will help you understand what to expect when taking PCP.

Report 1: A Trip of a Lifetime

Before my experience with PCP, I had only heard about it from my friends and never actually tried it. I was hesitant and had some doubts, but I decided to take a chance.

On the day of my experience, I was feeling a range of emotions, but I was mostly excited. When I took a small dose, I felt an immediate rush of energy and enthusiasm.

Initially, I experienced a sense of weightlessness and my senses seemed to be heightened. I began exploring my environment and took in very vivid and intense details that I never noticed before. I also felt like my physical and mental barriers had been removed and I could explore without caring about my limitations.

The most vivid, and intense, experience was the intense connection to the music that played in the background. Every musical note seemed to communicate with me directly.

I had an overall sense of connection to the world around me, but also felt that I had a certain detachment from reality. I was happy but at the same time felt like I was somehow removed from everyday life.

By the end of my experience, I felt a great sense of relaxation combined with a feeling of heightened optimism. This mood of optimism enabled me to work through challenges in my life which had previously seemed insurmountable.

Overall, my experience with PCP was positive and unbelievably intense. I felt a connection to my environment in ways that I had never experienced before, and I felt a sense of optimism that enabled me to work through problems and issues in life. I believe my experience with Phencyclidine was one of a kind and life-changing, both mentally and emotionally.

Report 2: Damaging Effects of Phencyclidine

Unfortunately, I had the dubious honor of personally experiencing the dangers of taking Phencyclidine (PCP). Though I was aware of the risks of indulging in the substance, I underestimated the harm it could cause.

Upon taking the drug, I was immediately struck with a wave of dizziness and nausea. Despite this uncomfortable feeling, I stoically soldiered on in my endeavor until worse symptoms started to present. Mental confusion and paranoia began to set in, leaving me feeling anxious and disoriented. I was struck with a dangerous sense of detachment from reality that I had never before experienced. My bodily senses began to fail, and I noticed intense and disorganized sensory input.

My hallucinations intensified as the hours wore on. I felt as though I was being watched, although I was aware this was not the case. The acuity of my senses heightened exponentially, making the smallest noises and sensations unbearable. My vision felt distorted and obstructed, making it difficult to pilot myself through any environment. Nightmares followed, taking me to places of extreme fear and dread.

It became increasingly hard to concentrate. Additionally, the overwhelming feelings of paranoia and fear continued to haunt me. I felt completely out of control and could not trust myself or my thoughts.

In the long-term, the physical and mental effects of the drug have not worn off. My anxiety is higher than ever, and I've been plagued with insomnia since the experience. The event has left me feeling constantly uneasy and terrified that I will succumb to the effects of the drug again.

I can unequivocally say, as a warning to anyone who may consider taking Phencyclidine, that it is not a substance to trifle with. The risks far outweigh any reward, physical or mental, and the effects are often more powerful and longer-lasting than anticipated.

Report 3: Experiencing Euphoria

I recently had my first experience with Phencyclidine (PCP), and it was a beautiful and profound experience. After carefully considering the risks that come with any use of an illicit substance, I decided I was ready to take the plunge and try some PCP.

I could immediately feel the effects within just a few minutes. The physical effects were mild; I was able to maintain an upright posture and remain in control of my body for the duration of my experience. I felt a deep sense of peace and calm, and I was pleasantly surprised at how quickly and dramatically I could change my mental state, even at such low doses.

The visual effects were by far the most profound. I could feel my mind opening up as time and space seemed to melt together and everything around me took on a vivid and beautiful quality. Colors and sounds seemed to blend together in a way I had never seen before and my thoughts and emotions flowed through me in an incredibly profound and positive way.

The most wonderful part of my experience was the intense feeling of euphoria and love I felt for everything around me. I could feel and appreciate the beauty of the world in a completely new way and I was filled with a deep and meaningful understanding of the complexities of the universe.

This experience was easily one of the most profound and transformative experiences of my life. I have a newfound appreciation and understanding of the world around us and the limitless possibilities that exist within our consciousness.

In conclusion, PCP, commonly known as Phencyclidine, is a potent and controversial substance with a rich history. Despite being developed in the 1950s as an intravenous anesthetic, it was eventually abandoned due to its unpredictable and dangerous side effects. Nevertheless, PCP found its way into the recreational drug market, where it continues to be used for its mind-altering effects. The historical background, chemical properties and effects, medical uses and abuses, and risks and side effects associated with PCP serve as a cautionary tale of the dangers of recreational drug use. Therefore, it is important to avoid PCP and drugs with unknown and potentially harmful effects and seek help for substance abuse issues.

Chapter Summary

- PCP (Phencyclidine) is a dissociative anesthetic developed in the 1950s for surgical procedures but was later abandoned due to its unpredictable side effects.
- PCP has a rich history and remains a controversial substance despite its dangers.

- PCP acts as a selective antagonist at the NMDA receptor and causes profound alterations in perception, including hallucinations and delusions.
- PCP is a white crystalline powder that can be taken orally or intravenously and can cause physiological effects like increased blood pressure and heart rate, muscle rigidity, and seizures.
- PCP was once used as a medical anesthetic but has since been banned in most countries and is now used as a recreational drug.
- The use of PCP as a recreational drug has been associated with negative outcomes such as hallucinations, delusions, and violence, as well as a high risk of addiction and long-term cognitive impairment.
- PCP is often used with other drugs, such as marijuana or alcohol, which can increase its potential for harm.
- PCP is associated with serious risks and side effects, including hallucinations, delusions, violence, seizures, muscle rigidity, and long-term cognitive impairment.

3

DXM (DEXTROMETHORPHAN)

As the sun sets on the horizon, a young man takes a deep breath and embraces the euphoria of his first DXM experience.

Dextromethorphan, also known as DXM is a **versatile substance** with a complex history and a range of effects. Originating as a cough suppressant in over-the-counter cold and flu medications, DXM has gained attention for its potential for abuse and illicit use as a recreational drug. This chapter will explore the historical background of DXM, its chemical properties and effects, medical uses and abuses, risks, and side effects, laws and regulations, and experience reports. Whether you're a medical professional, a curious individual, or someone looking for a comprehensive understanding of DXM, this chapter aims to provide an insightful and comprehensive guide to this multifaceted substance.

Historical Background

The history of Dextromethorphan, or DXM, dates back to the 1950s when it was first synthesized and developed as a cough suppressant. It was initially marketed in cough syrups and other over-the-counter cold and flu medications. As a result, its use as a cough suppressant became widespread. However, our society did not widely recognize its potential for abuse as a recreational drug until the late 20th century, when reports of individuals intentionally consuming large quantities of cough syrup for its psychoactive effects began to emerge.

In the decades since, DXM has become a widely abused substance, particularly among adolescents and young adults. In response to this trend, many countries have enacted laws and regulations restricting access to cough syrups containing DXM. In addition, some countries classify it as a controlled substance. Despite these measures, DXM abuse remains a significant public health concern. Consequently, it continues to be the subject of ongoing research and analysis.

Overall, the historical background of DXM reflects its journey from a simple cough suppressant to a widely abused recreational drug with a complex and evolving regulatory landscape. Its legacy continues to shape the perception and understanding of DXM. It serves as a reminder of the need for continued vigilance and awareness regarding substance abuse.

Chemical Properties and Effects

Dextromethorphan, or DXM, is a synthetic substance with unique chemical properties that give rise to its various effects. It is classified as a morphinan and acts as a non-competitive antagonist of the N-methyl-D-aspartate (NMDA) receptor in the central nervous system. This interaction results in the suppression of coughing and the alteration of perceptions and mood, making it both a cough suppressant and a recreational drug.

DXM is considered safe and effective for its intended use as a **cough suppressant**. However, higher doses can produce psychoactive effects, including euphoria, changes in perception, dissociation, and altered thinking and behavior. In addition, the specific effects of DXM can vary depending on the dose, individual tolerance, and other factors. Still, it is considered a dissociative substance.

It is important to note that the effects of DXM can be unpredictable, and its use carries significant risks, particularly when taken in large doses. Moreover, DXM can interact with other substances, including alcohol and drugs, increasing the risk of adverse effects and potentially life-threatening consequences.

In summary, the chemical properties of DXM play a significant role in determining its therapeutic and recreational effects. Therefore, understanding these properties is crucial to safely and effectively using this substance and avoiding its associated risks.

Medical Uses and Abuses

Dextromethorphan, or DXM, has both medical and recreational uses. It is widely available in over-the-counter cold and flu medications as a cough suppressant. In addition, it is used to treat symptoms of coughs, colds, and respiratory infections. DXM is considered safe and effective at recommended doses for its intended use.

However, the same properties that make DXM an effective cough suppressant make it attractive for abuse as a recreational drug. At high doses, DXM can produce psychoactive effects, including euphoria, changes in perception, and dissociation, leading some individuals to intentionally consume large quantities of cough syrup or other products containing DXM for its psychoactive effects. This use pattern is considered abuse and carries significant risks, including physical and psychological harm.

The abuse of DXM has been a growing concern in recent decades. It is particularly prevalent among adolescents and young adults. As a result, many countries have enacted laws and regulations to restrict access to cough syrups and other DXM products to address this trend, with some countries classifying it as a controlled substance.

In conclusion, while DXM has medical uses as a cough suppressant, its abuse as a recreational drug remains a significant public health concern. Therefore, it is important to understand the risks associated with DXM abuse and only use it as directed by a healthcare provider. Individuals should also be aware of the laws and regulations surrounding DXM and take steps to protect themselves and others from its harmful effects.

Risks and Side Effects

Dextromethorphan, or DXM, has both therapeutic and recreational uses. Still, its use carries significant risks and potential side effects, particularly in high doses. At low doses, DXM is considered safe for its intended use as a cough suppressant. Still, its abuse as a recreational drug can lead to various adverse effects, including physical and psychological harm.

Some of the most common side effects of DXM abuse include nausea and vomiting, dizziness and drowsiness, increased heart rate, and blood pressure, and changes in perception and thinking. High doses can also cause more severe side effects, such as hallucinations, agitation, confusion, seizures, and impaired coordination.

In addition to the physical side effects of DXM abuse, significant psychological risks are associated with its use. These risks can include feelings of anxiety and paranoia, depression, and long-term changes in mood and behavior. In addition, the dissociative effects of DXM can also lead to feelings of detachment and disorientation. They may increase the risk of accidents or harm oneself or others.

DXM can also interact with other substances, including alcohol and drugs, increasing the risk of adverse effects and potentially life-threatening consequences. It is also important to note that the long-term effects of DXM abuse are not well understood, and more research is needed to fully understand the risks and consequences associated with this pattern of use.

In conclusion, the risks and side effects associated with DXM use, both therapeutic and recreational, highlight the importance of using this substance only as directed by a healthcare provider and avoiding its abuse. Individuals should also be aware of the potential for adverse effects and take steps to minimize the risk of harm.

Laws and Regulations

The laws and regulations surrounding Dextromethorphan, or DXM, vary depending on the jurisdiction. Still, in many countries, it is subject to legal controls and restrictions to prevent abuse. This is due to the growing concern about the harmful effects associated with DXM abuse, particularly among adolescents and young adults.

In some countries, DXM is classified as a controlled substance. Its sale and distribution are subject to strict regulations, including age restrictions and prescription or over-the-counter availability requirements. In other countries, it is classified as a legal but regulated substance. Its sale and distribution may be subject to restrictions or limitations.

It is important to note that the laws and regulations surrounding DXM can change over time and vary from country to country. Therefore, individuals should be aware of the legal status of DXM in their jurisdiction and comply with all relevant laws and regulations regarding its use and abuse.

In addition to national laws and regulations, some international agreements may also play a role in controlling and regulating DXM. For example, the United Nations Convention on Psychotropic Substances requires signatory countries to control and regulate the production, distribution, and use of certain psychoactive substances, including DXM.

In conclusion, the laws and regulations surrounding DXM reflect the growing concern about its abuse and the associated risks. They are designed to minimize the harm associated with this substance. It is important for individuals to be aware of the legal status of DXM in their jurisdiction and to comply with all relevant laws and regulations regarding its use and abuse.

Experience Reports

Welcome to the experience reports section! Here, you can read first-hand accounts from people who have taken DXM (Dextromethorphan) and experienced its effects. Their stories provide valuable insight into the wide range of experiences that can occur when taking this substance – both positive and negative. In addition, these reports will help you understand what to expect when taking DXM.

Report 1: Reflecting on a Euphoric Experience with DXM

I recently had the opportunity to try DXM, and I can confidently say that it was an experience like no other. From the moment I took a dose of the substance, I found myself in a state of peace and tranquility that was both calming and uplifting.

I was amazed at the intense rush of euphoria that washed over me, feeling like I was disconnected from time and reality. All of my anxieties and worries seemed to float away, and in the place of them came feelings of joy and contentment. The colors around me looked brighter, and the music I was listening to sounded sweeter.

The visuals I experienced during this trip were also breathtaking. I saw vibrant colors and intricate patterns that filled my field of vision. For hours I felt like I could just sit and appreciate the beauty of it all.

In addition to the euphoria, DXM also gave me the mental clarity to more deeply reflect on life's big questions. I felt like I was able to make some extraordinary insights about my life and the world around me.

I was surprised to find that the effects of DXM lasted longer than I expected and that the comedown felt incredibly smooth. Despite the intensity of the experience, I felt well-rested, level headed, and energized when it was done.

In summary, DXM was an extremely positive experience for me. It not only left me feeling joyous and content, but I also gained a lot from it in terms of personal insights. I would highly recommend it to those looking to explore a unique state of consciousness.

Report 2: A Disaster of a DXM Trip

It all started out as harmless curiosity. I heard from a friend that DXM, an over-the counter cough syrup, could be used as a recreational drug to achieve a high. Naive as I was, I decided to try it for myself.

Regretfully, my DXM experience was nothing close to pleasant. As soon as I swallowed the syrup, feelings of anxiety and restlessness began to set in. My heart rate accelerated, my body became shaky, and my mind began to warp. I attempted to get comfortable by laying down, but my head began to race with streams of seemingly disconnected thoughts and ideas. I started to feel incredibly out of control and disoriented.

The physical effects of the DXM were awful—I felt extremely nauseous and dizzy, and I began vomiting uncontrollably. The sensation felt like an extended version of a car ride that's taken a wrong turn down a bumpy road. As the feeling intensified, I thought I was going to faint multiple times, and I was having difficulty thinking straight.

It would have been easy to just press pause, but the power of the drug seemed to cast a spell over me that I couldn't break. My trip seemed like it would never end, and every single thought felt like an eternity.

Now that it's been some time since my DXM experience, I'm glad it's finally over. I realized that day that DXM is not for me. It left me feeling empty, degraded, and scared. Overall, it was a disaster and a lesson learned.

Report 3: Riding the DXM Express

I recently decided to explore the depths of my psyche by experimenting with DXM (Dextromethorphan). My journey began with a measured dose of a DXM-containing cough syrup. As the viscous syrup trickled down my throat, I was beginning to get excited for the adventure that lay ahead.

The effects of the DXM started off slow and gentle, as I began to feel a pleasant tingling and warmth radiating throughout my body. Everything began to take on a dreamlike quality, as if I were living an alternate reality separate from the one I know. I felt immersed in an ever-shifting collage of colors and shapes, with my imagination running wild and my attention fixed firmly on the present.

My senses were heightened and I became acutely aware of the subtleties of sound and touch that usually go unnoticed. I experienced closeness to the people around me, regardless of whether or not I knew them and regardless of our differences. I had a sense of connectedness to all living beings.

The DXM had made me aware of the beauty in my own backyard and provided a new perspective on the world around me. I felt light, yet grounded; enlightened, yet grounded in the present moment.

The effects of the DXM continued to build and deepen as I became more and more introspective. I gained insight into myself and my relationships that I had not seen before. I finally felt at peace and ready to end my journey.

All in all, DXM was a unique and fascinating experience. It allowed me to explore my inner world and discover things about myself I had never known before. I am now more aware of my surroundings, my emotions, and my relationships with those around me. I have embraced the new perspective that DXM has given me and have come away from the experience feeling ready to take on life's challenges.

In conclusion, Dextromethorphan, or DXM, is a complex substance with a versatile range of effects, originating as a cough suppressant and later gaining attention as a recreational drug. This chapter provided an in-depth exploration of the historical background, chemical properties and effects, medical uses and abuses, risks, and side effects of DXM. It is important to understand the properties of this substance to safely and effectively use it and avoid its associated risks. While DXM has therapeutic uses, its abuse as a recreational drug remains a **significant public health concern**. Therefore, individuals should be aware of the laws and regulations surrounding it. Additionally, the use of DXM carries significant risks and potential side effects, and it is crucial only to use it as directed by a healthcare provider.

Chapter Summary

- DXM, also known as Dextromethorphan, is a versatile substance with a complex history, originally developed as a cough suppressant in over-the-counter cold and flu medications.
- Despite its widespread use as a cough suppressant, DXM has gained attention for its potential for abuse and illicit use as a recreational drug.
- DXM is classified as a morphinan and acts as a non-competitive antagonist of the N-methyl-D-aspartate (NMDA) receptor in the central nervous system, suppressing coughing and alterations in perceptions and mood.

- The effects of DXM can be unpredictable, and its use carries significant risks, particularly when taken in large doses or combined with other substances.
- DXM abuse has become a growing public health concern, particularly among adolescents and young adults, leading to regulations restricting access to cough syrups containing DXM in many countries.
- DXM has medical and recreational uses, with its abuse as a recreational drug carrying significant physical and psychological harm.
- The side effects of DXM abuse can range from nausea and vomiting, dizziness, increased heart rate, and blood pressure to more severe effects such as hallucinations, confusion, seizures, and impaired coordination.
- Understanding the risks and potential harm associated with DXM abuse is important for this substance's responsible and safe use.

4

NITROUS OXIDE (LAUGHING GAS)

A bright and sunny day at the park, with a rainbow of balloons floating in the sky, each one a unique and beautiful color.

Nitrous oxide, also known as laughing gas, balloons, chargers, hippie crack, or nangs, is a colorless and odorless gas that has captured people's interest for centuries. From its early use as a recreational substance in the 18th century to its widespread use in modern medicine, nitrous oxide has played a unique role in science, medicine, and recreation. This chapter will delve into the fascinating history of nitrous oxide, its chemical properties, effects, and medical applications, as well as the risks and side effects associated with its use. We will also explore the laws and regulations governing nitrous oxide and review some experience reports to better understand the recreational use of the gas. So join us on a journey through nitrous oxide's history, science, and cultural impact.

Historical Background

The history of nitrous oxide dates back to the late 18th century when English chemist and inventor Joseph Priestley first discovered the gas. However, the English chemist and natural philosopher Sir Humphry Davy recognized nitrous oxide's potential as a recreational substance. In the late 1700s and early 1800s, nitrous oxide parties, or "laughing gas parties," became popular in Europe and the United States. During these parties, people would inhale nitrous oxide for its euphoric and analgesic effects.

In addition to its recreational use, nitrous oxide has significantly impacted the field of medicine. In the late 19th century, American dentist Horace Wells used nitrous oxide as an anesthetic during dental procedures, marking the first recorded use of nitrous oxide in surgery. Over the next few decades, nitrous oxide became widely used in medicine, particularly in dentistry and surgery, due to its fast-acting and short-lived effects.

Today, nitrous oxide continues to be used as a medicinal and recreational substance. Yet, despite its long history, nitrous oxide still holds many **mysteries** and is the subject of ongoing scientific research and exploration.

Chemical Properties and Effects

Nitrous oxide, also known as dinitrogen oxide, is a colorless and odorless gas composed of nitrogen and oxygen. It has the chemical formula N_2O and is stored as a compressed liquid in steel cylinders. Nitrous oxide is a powerful anesthetic and a weak psychoactive substance with both physical and mental effects.

When inhaled, nitrous oxide rapidly diffuses into the bloodstream. It produces a range of effects, including increased feelings of euphoria, happiness, and excitement. It also reduces pain perception, creates a sense of detachment, and produces hallucinations and altered states of consciousness. The duration of nitrous oxide's effects depends on the dose and the method of administration but typically lasts between 5 and 20 minutes.

In addition to its psychoactive effects, nitrous oxide has several physical effects. It acts as a powerful analgesic, reducing pain perception and allowing surgical procedures without causing significant patient discomfort. Nitrous oxide also produces mild sedation and has a calming effect. It is useful for reducing anxiety and fear in medical and dental procedures.

It is important to note that nitrous oxide has a low therapeutic index, meaning that the difference between a therapeutic dose and a lethal dose is small. As a result, nitrous oxide must be used with caution and under the supervision of a trained medical professional.

Medical Uses and Abuses

Nitrous oxide has been used in medicine for over a century. It is widely recognized for its safe and effective use as an anesthetic. In the medical field, nitrous oxide is primarily used in dental procedures, such as extractions, fillings, and cleanings, as well as in surgical procedures to provide sedation and pain relief. In addition, nitrous oxide's fast-acting and short-lived effects make it an ideal choice for medical procedures requiring minimal sedation.

However, nitrous oxide also has a history of abuse, particularly in the recreational context. As a result, people inhale nitrous oxide for its psychoactive effects risk serious injury or death and long-term damage to their health. In addition, abusing nitrous oxide can cause a range of negative effects, including brain damage, nerve damage, and vitamin B12 deficiency, which can lead to peripheral neuropathy.

In addition, nitrous oxide abuse can lead to addiction, as users may become dependent on the drug to experience its psychoactive effects. The inhalation of nitrous oxide in large quantities can also lead to unconsciousness and, in some cases, death.

It is important to note that nitrous oxide is regulated in many countries and is only available for medical use with a prescription. Misuse or abuse of nitrous oxide is a serious crime and can result in severe legal consequences.

Risks and Side Effects

Like all substances, nitrous oxide has associated risks and side effects ranging from mild to severe. Some of the nitrous oxide's most common side effects include dizziness, headache, nausea, and vomiting. Individuals can sometimes manage these side effects with proper medical supervision. Still, in other cases, they may indicate a more serious problem.

One of the major risks associated with nitrous oxide is the risk of overdose. Overdose can occur when a person inhales too much nitrous oxide, leading to unconsciousness and, in some cases, death. Additionally, inhaling nitrous oxide can reduce the amount of oxygen in the body, leading to brain damage and other serious health problems.

Long-term nitrous oxide abuse can also lead to serious health problems, such as nerve damage, B12 deficiency, and peripheral neuropathy. Chronic abuse of nitrous oxide can also cause changes in the brain, leading to depression, anxiety, and other mental health problems.

In addition, nitrous oxide abuse can lead to addiction, as users may become dependent on the drug to experience its psychoactive effects. Withdrawal from nitrous oxide can lead to symptoms such as headaches, nausea, and fatigue and may require medical intervention.

It is important to use nitrous oxide only as directed by a trained medical professional and to never inhale nitrous oxide directly from a canister or other source. Misuse or abuse of nitrous oxide can result in serious health problems and should be avoided.

Laws and Regulations

Nitrous oxide is regulated by the government in many countries and is classified as a controlled substance in some jurisdictions. Therefore, the possession, sale, and distribution of nitrous oxide without a valid prescription are illegal and can result in serious legal consequences, including fines and imprisonment.

In the medical field, nitrous oxide is only used by trained healthcare professionals. It is subject to strict regulations regarding storage, handling, and administration. Medical facilities that use nitrous oxide must follow strict protocols to ensure the safe and proper use of the substance.

In recreational terms, nitrous oxide is often sold on the black market. It is commonly used at music festivals, parties, and other events. The sale and distribution of nitrous oxide for recreational purposes are illegal. They can result in severe legal consequences, including fines and imprisonment.

It is important to be aware of nitrous oxide laws and regulations and only use the substance as directed by a trained medical professional. Misuse or abuse of nitrous oxide can have serious legal consequences and should be avoided.

Experience Reports

Welcome to the experience reports section! Here, you can read first-hand accounts from people who have taken nitrous oxide and experienced its effects. Their stories provide valuable insight into the wide range of experiences that can occur when taking this substance – both positive and negative. In addition, these reports will help you understand what to expect when taking nitrous oxide.

. . .

Report 1: My Wild Journey with Nitrous Oxide

Recently, I decided to try something new - nitrous oxide. To be honest, I was a bit scared, but I am so glad I took the plunge!

I had a wonderful experience with nitrous oxide. Immediately upon inhaling it, my senses were heightened, and I felt an incredible sense of euphoria. All my worries were gone and I was free to explore my own world. The sounds were clearer, and the colors were brighter than ever before. I even felt a bit weightless, like I was floating on a cloud. It was an awe-inspiring experience, and the beautiful state of relaxation I was in lasted for hours.

Nitrous oxide made me feel incredibly peaceful and connected to the world around me.

Report 2: Dizzy and Disoriented

Today, I decided to try something new and experiment with nitrous oxide. It was a chore to get my hands on it, but I eventually found a source. After inhaling the nitrous oxide, I waited for the effects to be felt.

I can't say I was impressed. I immediately felt dizzy and out of it. I almost felt like I was going to faint, and I felt really off balance. My vision went blurry and I had trouble staying focused on the task at hand.

Maybe the worst part of it was the lack of control. I felt like a puppet, and the substance was the person pulling the strings. I had no control over my own body, and my mind felt fuzzier than usual.

Unfortunately, this experience did not go as planned. I have tried many substances and thought nitrous oxide was fine to try too. However, the lack of control and dizziness it caused was more than I bargained for. If anyone else out there was considering nitrous oxide, I do not recommend it.

Report 3: A Painless Experience

I had been curious about nitrous oxide for a long time, and I finally had the opportunity to experience it. I felt a bit apprehensive at first, but my worries were soon replaced by excitement and anticipation.

The setup was straightforward; I was in a comfortable chair with a breathing mask over my nose. It was easy to find a suitable breathing rhythm without giving it much thought. After a few minutes, I began to feel a wave of intense warmth and lightness of being. It felt like the room around me was spinning ever so slightly, and I felt a newfound sense of euphoria, like I was living in a dream.

The sensation didn't last very long; it felt like only a few minutes, but it was a great sensation. I felt energetic and relaxed, but also still in control. Physically, I felt pretty refreshed and relaxed.

I had no qualms with the experience, and I'd recommend it to anyone looking for a relaxing experience without the worry of any side effects. I feel lucky that I was able to experience nitrous oxide, and I'll be doing it again in the future.

In conclusion, nitrous oxide, also known as laughing gas, has played a unique role in science, medicine, and recreation for centuries. From its discovery by Joseph Priestly and recognition of its potential by Sir Humphry Davy to its widespread use in modern medicine, nitrous oxide has left a lasting impact on history. Its chemical properties, including its composition and effects, have made it a valuable tool in medicine as a powerful anesthetic. However, nitrous oxide also has risks and side effects associated with its use, especially when abused. Therefore, it is important to use nitrous oxide with caution and under the supervision of a trained professional, as misuse can lead to serious injury or death, addiction, and legal consequences.

Chapter Summary

- Nitrous oxide, also known as laughing gas, has a fascinating history dating back to the late 18th century.
- Nitrous oxide is a colorless and odorless gas composed of nitrogen and oxygen, used as a powerful anesthetic and weak psychoactive substance.
- When inhaled, nitrous oxide produces euphoria, happiness, reduced pain perception, detachment, hallucinations, and altered states of consciousness.
- Nitrous oxide has been used in medicine for over a century, mainly in dental and surgical procedures.
- Abusing nitrous oxide has serious risks, including brain damage, nerve damage, B12 deficiency, addiction, unconsciousness, and death.
- Misuse of nitrous oxide is a serious crime regulated by many countries.
- Nitrous oxide has common side effects, such as dizziness, headache, nausea, and vomiting.
- Chronic nitrous oxide abuse can lead to serious health problems, including nerve damage, B12 deficiency, peripheral neuropathy, depression, and anxiety.

5

MXE (METHOXETAMINE)

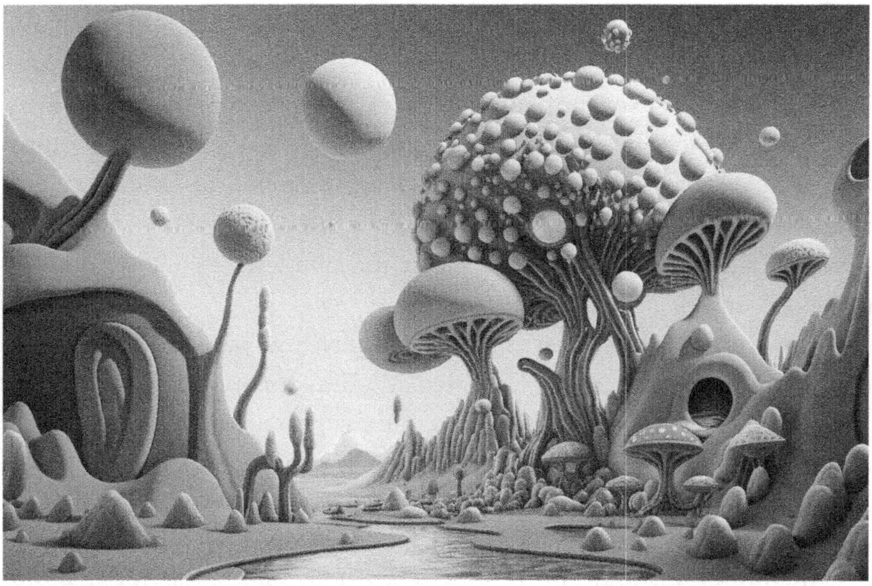

A surreal landscape where MXE molecules are the only inhabitants, floating in a sea of neon colors.

Methoxetamine, also known as MXE, M-Ket, Max, or Mexxy, is a novel psychoactive substance recently gaining popularity. Despite its growing use, many people need to become more familiar with MXE and its effects. This chapter will delve into the history and background of MXE, its chemical properties, and its various medical uses and abuses. We will also examine its use's risks and side effects and the laws and regulations governing it. Finally, we will provide a collection of experience reports from individuals who have used MXE, offering a glimpse into the personal experiences of those who have used the substance. Whether you are a researcher, healthcare professional, or simply curious about MXE, this chapter aims to provide you with a comprehensive understanding of this fascinating substance.

Historical Background

The historical background of Methoxetamine, commonly known as MXE, is relatively brief. MXE was first synthesized in the **early 2010s** and quickly gained popularity as a recreational drug due to its supposed similar effects to the dissociative anesthetic ketamine. It was first sold online as a research chemical and marketed as a legal alternative to ketamine.

However, due to concerns over its safety and potential for abuse, many countries have since classified MXE as a controlled substance. Despite this, it remains readily available for purchase on the dark web and in some countries where it is not yet banned.

MXE has gained significant attention from the media and law enforcement, with reports of serious adverse events associated with its use, including hospitalizations and death. This highlights the need for further research into the effects and potential dangers of MXE and other similar substances.

It is important to note that the limited research available on MXE suggests that its long-term effects still need to be fully understood and may be much more potent and dangerous than originally thought. As a result, individuals are highly recommended to avoid using MXE and seek safer, more researched alternatives for recreational or therapeutic purposes.

Chemical Properties and Effects

MXE, or Methoxetamine, is a synthetic compound classified as a dissociative anesthetic. This means that it produces a state of detachment or dissociation from one's surroundings and physical sensations, leading to a feeling of floating and disconnection.

MXE has a chemical structure similar to ketamine. Its effects are thought to be caused by its actions on the glutamate and dopamine neurotransmitter systems. It is also believed to bind to the NMDA receptor in the brain, leading to its anesthetic and dissociative effects.

The effects of MXE can last anywhere from a few hours to several days and include altered perception of time and space, changes in body sensations, feelings of euphoria and detachment, and hallucinations. It can also lead to confusion, agitation, and intense hallucinations, known as a "bad trip." At high doses, MXE can cause unconsciousness, which can be dangerous and lead to accidental injury.

It is important to note that due to the limited research on MXE, its effects, and potency must be better understood. The substance can be unpredictable and potentially dangerous, and we should approach its use cautiously. Suppose you are considering using MXE or any other psychoactive substance. In that case, it is recommended that you speak with a healthcare professional and carefully research its effects, risks, and potential dangers.

Medical Uses and Abuses

MXE, or Methoxetamine, has not been approved for medical use and is not recognized as a safe or effective treatment for any medical condition. However, it has been used recreationally and abused for its dissociative and hallucinogenic effects.

Despite its lack of medical approval, some individuals have claimed that MXE has helped them with depression, anxiety, and other mental health conditions. However, these claims have not been scientifically proven, and using MXE for medical purposes is not recommended.

The recreational use of MXE is associated with several potential risks, including psychological dependence, tolerance, and withdrawal. Additionally, using MXE can lead to serious adverse events, such as accidental injury, overdose, and death.

MXE abuse has also been associated with negative physical and psychological effects, including confusion, agitation, panic attacks, delusions, and hallucinations. It can also lead to dangerous health consequences like cardiac and respiratory problems, seizures, and kidney failure.

It is important to note that using MXE, whether for medical or recreational purposes, carries significant risks and potential dangers. Therefore, if you are considering using MXE or any other psychoactive substance, it is recommended that you speak with a healthcare professional and carefully consider the potential risks and benefits. Abusing any substance can have serious consequences for your health and well-being and should be avoided.

Risks and Side Effects

MXE, or Methoxetamine, is associated with several potential risks and side effects. These can range from short-term effects, such as confusion and agitation, to more serious and long-lasting health problems. Some of the common short-term side effects of MXE include:

- Altered perception of time and space
- Changes in body sensations
- Confusion and disorientation
- Nausea and vomiting
- Agitation and panic attacks
- Hallucinations

The long-term effects of MXE use are not well understood, but some potential consequences include the following:

- Psychological dependence and tolerance
- Withdrawal symptoms, including anxiety and depression
- Cognitive impairments, such as memory problems and difficulty concentrating
- Respiratory and cardiovascular problems
- Kidney and liver damage
- Seizures and other neurological problems

It is important to note that using MXE can be **unpredictable** and lead to serious adverse events, including overdose and death. Therefore, if you are experiencing any negative effects from using MXE, it is recommended that you seek medical attention immediately.

Furthermore, MXE is not regulated, and its quality and purity can vary greatly, making it difficult to know what you are taking. This increases the risk of accidental overdose and serious health problems.

In conclusion, MXE, like any other psychoactive substance, carries significant risks and potential dangers. If you are considering using MXE, it is recommended that you speak with a healthcare professional and carefully weigh the potential risks and benefits. It is always safer to avoid using untested and unregulated substances.

Laws and Regulations

MXE, or Methoxetamine, is a relatively new substance whose legal status varies from country to country. In many countries, including the United States, the United Kingdom, and Australia, MXE is classified as a controlled substance, and its sale, possession, and use are illegal.

In the United States, MXE is classified as a Schedule III controlled substance, which means that it is considered to have a high potential for abuse and dependence and is illegal to manufacture, distribute, or possess.

MXE is classified as a Class B drug in the United Kingdom, and its sale, possession, and use are illegal. The penalties for violating the law regarding MXE can be severe, including fines and imprisonment.

In Australia, MXE is classified as a Schedule 9 substance, which means that it has a high potential for abuse and dependence and is illegal to manufacture, distribute, or possess.

It is important to note that the legal status of MXE can change rapidly, and it is recommended that you check the current laws and regulations in your jurisdiction before using or possessing the substance.

In conclusion, using and possessing MXE is illegal in many countries, and the penalties for violating the law can be severe. Therefore, if you are considering using MXE or any other psychoactive substance, it is recommended that you research the current laws and regulations in your jurisdiction and consult with a legal professional if necessary.

Experience Reports

Welcome to the experience reports section! Here, you can read first-hand accounts from people who have taken MXE (Methoxetamine) and experienced its effects. Their stories provide valuable insight into the wide range of experiences that can occur when taking this substance – both positive and negative. In addition, these reports will help you understand what to expect when taking MXE.

Report 1: A Blissful and Enlightening Journey with MXE

I recently had the experience of trying MXE, also known as methoxetamine. I settled into a comfortable chair, making sure I had all that I needed close by in case I wanted to move or shift my position.

The effects of the MXE slowly started to come on within about thirty minutes of taking it. I felt a noticeable shift in my mood, in addition to slight tingling sensations throughout my body. As the MXE fully began to take effect, I felt myself being taken away on an incredibly blissful journey. I experienced complete relaxation, a feeling of total euphoria, and a deep and profound connection to my inner being.

The MXE seemed to stimulate a deep contemplation within me, as I contemplated my life and the meaning of existence. I felt a true sense of peace and understanding, as I was able to look at life from a different perspective. A message of unconditional love resonated with me at the most profound levels.

As I moved towards the end of the MXE experience, I felt open and connected to my surroundings once more. I felt a deep sense of joy and freedom, as my insight and understanding of life had been completely altered for the better. Overall, the experience with MXE was incredibly peaceful and enlightening, and I am so grateful for the opportunity to have experienced it.

Report 2: My Horrible MXE Experience

I recently tried MXE, also known as methoxetamine, for the first time and it was an experience I never want to repeat. Going into it, I heard a lot of positive things about the drug, so I figured it sounded relatively harmless. Boy, was I wrong.

I took the drug and at first, I felt a sense of euphoria. However, the good feelings quickly vanished and were replaced with an intense rush of anxiety and paranoia. I felt an overwhelming sense of dread and my entire body felt completely overwhelmed. I went from feeling calm and relaxed to feeling total terror in a matter of moments.

The physical effects of MXE were equally unpleasant. I felt nauseous and lightheaded and I couldn't keep my balance. My vision blurred and every sound I heard felt like a million knives stabbing through my brain. My heart raced and felt like it was about ready to stop and I thought for sure I was going to go into cardiac arrest.

After what felt like an eternity, the effects of the drug finally wore off and I was left exhausted, drained and feeling worse than I ever had in my entire life. The hangover lasted for days afterwards, making it that much worse.

I regret ever trying MXE and would never recommend it to anyone. Avoid it at all costs!

Report 3: An Unexpected and Positive Psychedelic Experience with MXE

I recently had the opportunity to experiment with MXE for the first time and it was an unforgettable experience. After ingesting it, I began to feel a deep sense of relaxation, both mentally and physically. I felt like I was in a trance and could hear music in the background as if it was playing from within my own self.

As time passed, I began to drift away to a place of beautiful imagery. Colors and shapes began to appear, spinning and dancing in front of me. I felt as if I had transcended to a higher level of consciousness and was able to see things in a completely new way.

The effects of MXE lasted for several hours before slowly beginning to subside. As I gradually returned to my everyday consciousness, I was left with an overwhelmingly positive feeling. I felt inspired and energized, with a newfound clarity of mind.

For someone looking for a light, positive psychedelic experience, I would certainly recommend giving MXE a try. It opened my mind up to a fascinating new world of possibilities, and I'm grateful I had the chance to explore it.

In conclusion, Methoxetamine, also known as MXE, is a novel psychoactive substance that has recently gained popularity but remains largely misunderstood. Despite its growing use, it is important to understand its history, chemical properties, and medical uses and abuses. MXE is classified as a dissociative anesthetic and has similar effects to ketamine. Still, its long-term effects are not fully understood and may be much more potent and dangerous. The recreational use of MXE is associated with several potential risks and side effects, including psychological dependence, tolerance, withdrawal, and serious adverse events like accidental injury, overdose, and death. Therefore, individuals are highly recommended to avoid using MXE and seek safer, more researched alternatives. Before using any psychoactive substance, it is important to speak with a healthcare professional and carefully consider the potential risks and benefits.

Chapter Summary

- MXE, also known as Methoxetamine, is a synthetic dissociative anesthetic that has gained popularity as a recreational drug.
- MXE was first synthesized in the early 2010s and was marketed as a legal alternative to ketamine.
- MXE has been classified as a controlled substance in many countries due to concerns over its safety and potential for abuse.
- MXE produces a state of detachment or dissociation from one's surroundings and physical sensations, leading to a feeling of floating and disconnection.
- MXE has not been approved for medical use, and its potential risks and side effects still need to be fully understood.
- The recreational use of MXE is associated with several potential risks, including psychological dependence, tolerance, and withdrawal.
- Abusing MXE can lead to serious adverse events, such as accidental injury, overdose, and death, as well as negative physical and psychological effects.

- Using MXE carries significant risks and potential dangers and should be approached with caution. Speak with a healthcare professional before using MXE or any other psychoactive substance.

6

3-MEO-PCP

A young man lies in a state of euphoria, surrounded by clouds of pcp-induced smoke, while a rainbow of colors dance around him.

In recent years, 3-MeO-PCP, also known as 3-Methoxy-Phencyclidine, has gained attention due to its growing popularity as a recreational drug. This chapter will overview the historical background, chemical properties, and effects of 3-MeO-PCP. Additionally, we will delve into this substance's medical uses and abuses and the risks and side effects associated with its consumption. This chapter will also discuss the laws and regulations surrounding 3-MeO-PCP and personal accounts or experience reports from those who have used the substance. By the end of this chapter, readers will have a comprehensive understanding of 3-MeO-PCP and its impact on society.

Historical Background

The historical background of 3-MeO-PCP dates back to the late 1950s when phencyclidine (PCP) was first synthesized and developed as an anesthetic. However, due to its unpredictable and often dangerous side effects, it was discontinued for medical use. Instead, it gained notoriety as a recreational drug in the 1970s and 1980s.

In recent years, new derivatives of PCP have emerged, including 3-MeO-PCP. This substance has gained popularity among drug users for its hallucinogenic effects, which are reported to be similar to those of PCP, but with a shorter duration. Although 3-MeO-PCP is not as widely known as other recreational drugs, its use is growing. Unfortunately, it has been associated with many negative consequences, including overdoses and hospitalizations.

Despite its relatively short history as a recreational drug, 3-MeO-PCP has already established a reputation for being dangerous and unpredictable. As the use of this substance continues to increase, it is important to understand its effects and potential consequences, as well as its historical background and development.

Chemical Properties and Effects

3-MeO-PCP is a synthetic compound that belongs to a class of drugs known as dissociative anesthetics. It disrupts communication between the brain and body, resulting in feelings of detachment, altered perception, and, sometimes, hallucinations.

The chemical structure of 3-MeO-PCP is similar to that of PCP but with a few key differences. These modifications to the molecular structure have been reported to have different effects. For example, some users describe 3-MeO-PCP as having a more euphoric and less intense dissociative experience than PCP.

The effects of 3-MeO-PCP vary depending on the dose, method of administration, and individual characteristics of the user. Low doses can produce mild stimulation, euphoria, and mild hallucinations. Higher doses can cause intense **dissociation**, sedation, and significant alterations in perception and thinking.

It is important to note that 3-MeO-PCP is a potent and highly unpredictable substance. As a result, its effects can vary widely, even among experienced users. This unpredictability and the drug's tendency to produce intense and sometimes frightening experiences make it a particularly dangerous substance to use recreationally.

Medical Uses and Abuses

3-MeO-PCP was never approved for medical use and had no accepted medical applications. Despite this, some people have sought out the drug for its dissociative effects and potential to produce hallucinations.

The recreational use of 3-MeO-PCP has become increasingly popular recently, particularly among those seeking a more intense or hallucinogenic experience. The drug is typically consumed orally, although some users may snort or inject it.

The abuse of 3-MeO-PCP can have serious consequences, including overdose, hospitalization, and long-term health problems. In addition, the effects of the drug can be highly unpredictable, even among experienced users. They can include extreme anxiety, agitation, and violent behavior.

In addition to its potential to cause physical harm, 3-MeO-PCP can also have significant psychological effects, including hallucinations, delusions, and paranoia. These effects can be particularly dangerous for individuals with pre-existing mental health conditions, as the drug can exacerbate their symptoms and lead to a more severe reaction.

Given the serious risks associated with the abuse of 3-MeO-PCP, individuals need to be aware of the dangers of using this substance and avoid it altogether. Furthermore, healthcare providers and substance abuse specialists should be prepared to address the issue of 3-MeO-PCP abuse and its consequences as the use of this drug continue to increase.

Risks and Side Effects

Using 3-MeO-PCP can carry significant risks and side effects in the short and long term. Some of the most common side effects include:

- **Dissociation:** This is the most pronounced effect of 3-MeO-PCP and can result in feelings of detachment from reality, altered perceptions of time and space, and, sometimes, hallucinations.
- **Agitation and anxiety:** The drug can cause intense agitation and anxiety, particularly at higher doses, leading to dangerous and unpredictable behavior.
- **Physical symptoms:** 3-MeO-PCP can cause many physical symptoms, including high blood pressure, heart palpitations, muscle rigidity, and seizures.
- **Cognitive impairment:** The drug can cause significant cognitive impairment, including confusion, memory loss, and difficulty with problem-solving and decision-making.
- **Overdose:** 3-MeO-PCP is a potent drug; even a small overdose can result in serious consequences, including coma and death.

In addition to the short-term risks, 3-MeO-PCP use can lead to long-term health problems. For example, frequent drug use can cause tolerance, increase the risk of overdose, and cause lasting damage to the brain and other organs.

Given these serious risks and side effects, individuals need to be aware of the dangers of using 3-MeO-PCP and avoid it altogether. Suppose you or someone you know is struggling with substance abuse. In that case, it is important to seek help from a healthcare provider or substance abuse specialist.

Laws and Regulations

3-MeO-PCP is a controlled substance in many countries, and its possession, sale, and distribution are strictly regulated. The same laws and regulations surrounding 3-MeO-PCP vary from country to country. Still, in general, the drug is classified as a Schedule I or Schedule II controlled substance, meaning that it is considered to have a high potential for abuse and no accepted medical use.

In the United States, 3-MeO-PCP is classified as a Schedule I controlled substance under the Controlled Substances Act. This means that it is illegal to manufacture, distribute, or possess the drug, and penalties for violating these laws can be severe, including fines and imprisonment.

In the European Union, 3-MeO-PCP is regulated by the European Monitoring Centre for Drugs and Drug Addiction (EMCDDA) and is classified as a controlled substance in most member states.

Given the strict laws and regulations surrounding 3-MeO-PCP, individuals must be aware of the legal consequences of using and possessing the drug. Suppose you have questions about your area's laws and regulations surrounding 3-MeO-PCP. In that case, it is best to consult a knowledgeable attorney or legal expert.

Experience Reports

Welcome to the experience reports section! Here, you can read first-hand accounts from people who have taken 3-MeO-PCP and experienced its effects. Their stories provide valuable insight into the wide range of experiences that can occur when taking this substance – both positive and negative. In addition, these reports will help you understand what to expect when taking 3-MeO-PCP.

Report 1: The Most Amazing Insight into Reality

When I decided to try 3-MeO-PCP, I had no idea how much I'd be rewarded. This trip was truly remarkable.

The experience began slowly - I felt visual distortions that slowly became more intense. I experienced more depth and variety of color than ever before and I felt that my surroundings had taken on an entirely new life. I found myself smiling more and more as the effects became stronger, like I was in a secret and sacred place that no one else knew.

The peak of the experience turned out to be even more remarkable. I felt incredibly connected with my inner self and the universe around me. I felt like I understood life in a way I could never have imagined. Everything I saw and experienced felt so real, so perfect, and it made me feel so much love for the people and things around me. I came away from the experience with a heightened understanding of life and the way I wanted to live it.

Overall, my experience with 3-MeO-PCP was truly remarkable. I feel that it has opened me up to new ways of looking at the world and I'm grateful for the insights I've gained through this amazing experience.

Report 2: No More

. . .

I can't believe I did it. I'm so disappointed in myself that I actually had the audacity to try 3-MeO-PCP. I had heard that it was supposed to be a powerful psychedelic and a dissociative, but I had no idea just how intense the experience would be.

I had thought that I was prepared to handle its psychedelic effects, but boy was I wrong. As soon as I took it, I immediately felt the urge to vomit, and my heart started pounding out of my chest. I felt like I was having an out-of-body experience as my sense of reality slowly started to slip away.

I attempted to hold onto my sense of reality with every ounce of strength that I had, but it was too late. I was mentally, emotionally, and physically overloaded by the 3-MeO-PCP. I had the worst trip of my life, and my mental state was in shambles when it eventually began to wear off.

Never again will I make the mistake of taking 3-MeO-PCP. This experience was far too intense, and I'm not sure if I'll ever be able to fully recover from it. My advice to anyone considering trying 3-MeO-PCP is to think twice before doing so. It is an incredibly powerful substance and should not be taken lightly.

Report 3: My Trippy Trip to the Third Dimension with 3-MeO-PCP

I recently had the pleasure of exploring the otherworldly splendor that 3-MeO-PCP had to offer. After carefully reading up and researching the potential effects, I felt prepared to embark on my own psychedelic journey.

I snuck away to a quiet and comfortable area, where I could feel safe and secure while I "took off". This was paramount to the experience, as I felt it allowed myself to be at ease and not be opposed to the expedition I was about to embark on.

I took the 3-MeO-PCP in the form of an insufflated dose to get the most intense effects, and the results were beyond my expectations. A wave of euphoria engulfed me, and I had a greater connection to the world around me. I was still aware of what was going on, but visions and insights presented themselves to me, and I felt as though I was exploring a brightly colored, fully-immersive world, so vivid and intense it verged on the unreal. I was transported to the beyond, into a world that I would never forget.

When the effects started to wear away, I relaxed, feeling mesmerized but contented. I returned to the real world with newfound facial expressions, a spiritual connection and a greater appreciation for life. I had certainly explored a new dimension, one that I plan to visit again and again.

In conclusion, 3-MeO-PCP, a synthetic compound, and dissociative anesthetic, has gained popularity as a recreational drug due to its hallucinogenic effects. However, its use has been associated with negative consequences such as overdoses, hospitalizations, and long-term health problems. This substance is highly unpredictable, and its effects can vary widely, even among experienced users, leading to potentially dangerous situations. Furthermore, the drug has no accepted medical applications. It can cause significant psychological harm, particularly in individuals with pre-existing mental health conditions. Individuals must be aware of the dangers of using 3-MeO-PCP and avoid it altogether. Healthcare providers and substance abuse specialists must be prepared to address the issue as its use continues to increase.

Chapter Summary

- 3-MeO-PCP, also known as 3-Methoxy-Phencyclidine, is a growingly popular recreational drug with hallucinogenic effects.

- The historical background of 3-MeO-PCP dates back to the late 1950s when it was first synthesized and developed as an anesthetic but discontinued due to its dangerous side effects.
- 3-MeO-PCP is a synthetic compound that belongs to a class of drugs known as dissociative anesthetics, disrupting communication between the brain and body.
- The effects of 3-MeO-PCP vary with dose, method of administration, and individual characteristics and can range from mild stimulation to intense dissociation and alterations in perception.
- 3-MeO-PCP was never approved for medical use and had no accepted medical applications.
- The abuse of 3-MeO-PCP can result in serious consequences such as overdose, hospitalization, long-term health problems, and worsening mental health conditions.
- The side effects of 3-MeO-PCP use include dissociation, agitation and anxiety, physical symptoms, cognitive impairment, and overdose.
- 3-MeO-PCP use carries significant short and long-term risks, and individuals need to be aware of its dangers and avoid it. In addition, healthcare providers and substance abuse specialists must be prepared to address the issue of its abuse.

7

SALVIA DIVINORUM

A woman enters a dark cave, illuminated only by the moonlight that streams from a hole in the ceiling. Inside, she finds a strange plant with purple flowers - Salvia Divinorum.

Salvia divinorum, also known as Maria Pastora, Sally-D, or simply Salvia, is a powerful psychoactive plant used for centuries for its mind-altering effects. With its growing popularity, it's essential to understand the historical background, chemical properties, medical uses, risks, and laws surrounding this plant. In this chapter, we will dive into salvia divinorum and explore its impact on human consciousness. From its ancient origins to its modern-day use, we'll cover everything you need to know about this intriguing substance. We'll also touch on the potential medical benefits and the possible dangers associated with its use. Lastly, we'll examine salvia's legal landscape and share some personal experience reports to give you a glimpse into what to expect. Whether you're a seasoned salvia user or just curious about this unique plant, this chapter offers something.

Historical Background

The historical background of salvia divinorum dates back to the indigenous people of Mexico, who have used the plant for its psychoactive effects for centuries. The Mazatec indigenous tribe, who lived in the Oaxaca region of Mexico, were the first to discover the plant's mystical properties. They used salvia divinorum in traditional shamanic healing ceremonies, considered a powerful tool for inducing spiritual visions.

The plant was largely unknown to the Western world until the **mid-20th century** when anthropologists and botanists first learned of its use by the Mazatec tribe. It was later introduced to the Western world in the 1950s and 1960s, quickly gaining popularity among psychonauts and those seeking a new and unique mind-altering experience.

In the early 2000s, salvia divinorum started gaining notoriety as a recreational drug. Its popularity spread rapidly, especially among young people. This led to increased concerns about its safety and potential risks, which prompted many countries to regulate its use and sale.

Today, salvia divinorum continues to be a subject of fascination, with new research being conducted on its effects and potential medical uses. Despite the controversies surrounding its use, the plant remains an important part of cultural and spiritual traditions for many people. Moreover, its historical background is a testament to its lasting impact on human consciousness.

Chemical Properties and Effects

The active ingredient in salvia divinorum is salvinorin A, a highly potent and selective kappa opioid receptor agonist. This means that when ingested, salvinorin A binds to the kappa opioid receptors in the brain and produces a range of effects, including alterations in perception, thought, mood, and consciousness.

The effects of salvia divinorum are highly variable. They can range from mild to intense, depending on the dose and method of ingestion. When taken in low doses, the plant can produce mild effects such as a mild buzz or feeling of relaxation. However, when taken in higher doses, the effects can be much more intense and can last anywhere from 15 minutes to an hour.

Some of the most commonly reported effects of salvia divinorum include changes in perception, such as visual and auditory hallucinations, feelings of detachment from the body, and intense euphoria or anxiety. The plant can also cause alterations in thought patterns, leading to a profound change in one's perspective or worldview.

It is important to note that the effects of salvia divinorum can be highly unpredictable and can vary greatly from person to person. This variability makes it difficult to predict the exact nature and intensity of the effects. It underscores the importance of caution and responsible use when experimenting with this substance.

Medical Uses and Abuses

Salvia divinorum has been studied for its potential medical uses. Although the research is still in its early stages, there is evidence to suggest that the plant may have therapeutic potential. Some of the most promising areas of research include the treatment of addiction, depression, and pain.

For example, studies have shown that salvia divinorum has the potential as a treatment for opioid addiction. The plant's active ingredient, salvinorin A, has been shown to reduce cravings and withdrawal symptoms in people addicted to opioids. Additionally, the plant's ability to induce feelings of detachment from the body and alter perception may make it useful as a therapeutic tool for people struggling with depression and anxiety.

Despite these potential benefits, it is important to note that salvia divinorum for medicinal purposes still needs to be better understood. More research is needed to understand its effects and potential risks fully.

On the other hand, the plant can also be abused, particularly as a recreational drug. Its intense and highly variable effects can lead to dangerous and unpredictable outcomes, particularly when taken in large doses or combined with other substances. Additionally, the plant's potency and rapid onset of action can lead to accidental overdoses, which can be dangerous and even life-threatening.

Therefore, it is crucial to approach salvia divinorum with **caution** and use it only under the supervision of a trained medical professional. Furthermore, individuals should carefully balance the plant's potential benefits against its potential risks, and the latest scientific evidence should guide its use.

Risks and Side Effects

Salvia divinorum is a potent psychoactive plant that can significantly affect the mind and body. While the plant has been used for centuries for its medicinal and spiritual properties, it also carries some risks and side effects that individuals must consider.

Some of the most common side effects of salvia divinorum include nausea, dizziness, headache, loss of coordination, and impaired judgment. Additionally, the plant's intense and unpredictable effects can be dangerous, particularly for people with a history of mental health problems, as it can trigger psychotic symptoms or exacerbate existing conditions.

In some cases, using salvia divinorum can lead to accidents or dangerous behaviors, such as driving while under the influence or engaging in dangerous activities while under the influence. This highlights the importance of using caution and avoiding heavy operating machinery or engaging in any other activities that require alertness and good judgment while under the influence of the plant.

Additionally, the plant's potency and rapid onset of action can lead to accidental overdoses, which can be dangerous and even life-threatening. Therefore, it is important to be aware of the potential risks associated with using salvia divinorum and approach its use with caution, particularly for people who are sensitive to its effects or have a history of mental health problems.

In conclusion, while salvia divinorum has potential therapeutic benefits, it is important to be aware of its potential risks and side effects. Therefore, its use should be guided by the latest scientific evidence and only under the supervision of a trained medical professional.

Laws and Regulations

The legal status of salvia divinorum varies widely across the world, with some countries regulating or outright banning its use. In contrast, others allow it to be sold and consumed freely.

In the United States, salvia divinorum is a controlled substance in several states, meaning its sale, possession, and use are restricted by law. However, the plant is unregulated in other states and can be purchased and used freely. At the federal level, salvia divinorum is not a controlled substance. Still, the Drug Enforcement Administration (DEA) has listed it as a "Drug of Concern."

In Europe, the legal status of salvia divinorum also varies widely from country to country. Some countries, such as the United Kingdom, regulate its use, while others, such as the Netherlands, allow its use without restriction.

It is important to note that the legal status of salvia divinorum can change rapidly and is subject to various interpretations and regulations. Therefore, it is essential to stay informed about the latest laws and regulations regarding its use in your specific jurisdiction and to be mindful of the potential legal consequences of using this substance.

In conclusion, the legal status of the salvia divinorum is complex and rapidly evolving, with different countries and jurisdictions adopting different approaches to regulating its use. Therefore, before using this substance, it is important to be aware of the laws and regulations in your area and to use it only in compliance with the latest legal requirements.

Experience Reports

Welcome to the experience reports section! Here, you can read first-hand accounts from people who have taken salvia divinorum and experienced its effects. Their stories provide valuable insight into the wide range of experiences that can occur when taking this substance – both positive and negative. In addition, these reports will help you understand what to expect when taking salvia divinorum.

Report 1: My Amazing Salvia Divinorum Trip

I recently had the pleasure of trying salvia divinorum for the first time, and the experience was nothing short of remarkable. The intense physical sensations and changes in my perception of reality were the most unusual sensations I've ever experienced.

The trip began with me ingesting some of the salvia and sitting down in a comfortable chair, prepared for what I was about to experience. At first, I felt a faint tingling throughout my whole body, and a feeling of lightheadedness as though I was entering a daze. The sensations progressed rapidly, and before I knew it, I was falling through a never-ending tunnel. I felt a strange sense of freedom and detachment, which seemed to be an upside-down form of meditation.

As I ventured further, I noticed my environment had changed, and I was now walking through surreal forests, lush meadows, and curious structures made from random objects. Colors seemed to glow from everything, and the music I heard was changing from moment to moment, morphing from classical tunes to different genres of electronic music. It was unlike anything I'd ever heard before.

At some point, I became aware of a small creature standing in front of me. It looked like an elf or a fairy, but I couldn't really tell what it was. I felt an immediate connection with it, as if we had known each other for centuries. We talked for a while and I asked it questions about the universe. It gave me advice and shared its wisdom with me. After a while, I began to feel as though I was slowly coming back to Earth, or at least a version of it.

My salvia divinorum experience was an incredible one. I'm so glad that I had the chance to explore this amazing plant medicine and uncover its many mysteries. It was both a fascinating and enlightening journey, which taught me a lot about life and the universe. Without a doubt, it was one of the most incredible trips of my lifetime.

Report 2: A Destructive Experience

Recently, I decided to experiment with the powerful natural psychedelic known as salvia divinorum. After doing some research, I felt perhaps I had a grasp of what I was getting into. However, nothing could have prepared me for this utterly terrifying experience.

From the moment I ingested the small amount of salvia divinorum, I felt a strange and overwhelming sense of dread. Over the course of the next hour, my anxiety only built as things around me seemed to become warped and distorted. Colors were more intense and I experienced a frightening sense of detachment from reality. As if I observing a bizarre dream, I was unable to physically control my own limbs or any movements.

I had heard that this experience can be overwhelming and even uncomfortable, but I had no idea it could be so destructive. I felt as though my mind was being broken down and split into different pieces. I could no longer trust any of my senses, and I was in a constant state of fear and confusion.

The effects did eventually wear off and I am now safely back to reality. I feel as though I am still trying to make sense of everything that happened, but in hindsight, I can confidently say that salvia divinorum is not for me. I highly suggest that others abstain from experimenting with this drug, as the experience was far from pleasant.

Report 3: A Tale of Illuminating Inspiration

I recently had the opportunity to try out salvia Divinorum as a means of achieving personal enlightenment. It's a type of sage that is native to Mexico and is known to enhance mental clarity and inspire creativity.

I had heard rumors of people claiming to experience tremendous spiritual journeys under its influence, so I was extremely excited to see what it could do for me. The afternoon of my experiment, I set up my room with some gentle music, candles, and plenty of pillows for ultimate comfort.

When I finally took the plunge and tried it for the first time, I was amazed to find out how powerful this leaf was! I experienced a surge of strange visuals and audible hallucinations that lasted for about fifteen minutes. During this span of time, I explored never-before-imagined realms and received great insight into some of the more complicated issues in my life.

The result of my experience was undeniably positive. I was able to gain insight into areas of my life that had been difficult to pinpoint and make sense of before. Additionally, I felt inspired and became motivated to pursue the dreams and goals that I had been neglecting for some time.

Overall, using salvia Divinorum left me feeling deeply inspired and loved. It was a wonderful opportunity to explore my innermost thoughts, and a great reminder that there is much more to life than the mundane. I couldn't speak more highly of this experience, and would recommend it to anyone looking for an enlightening journey.

In conclusion, salvia divinorum is a powerful psychoactive plant with a rich historical background that has been used for centuries by the indigenous people of Mexico for its mind-altering effects. Its active ingredient, salvinorin A, binds to the kappa opioid receptors in the brain and produces a range of effects, including alterations in perception, thought, mood, and consciousness. While research is ongoing, there is evidence to suggest that salvia divinorum may have therapeutic potential, particularly in treating addiction, depression, and pain. However, it is important to approach its use with caution and to consider its potential risks and side effects, including nausea, dizziness, impaired judgment, and unpredictable effects. Overall, it is crucial to use salvia divinorum only under the supervision of a trained medical professional and to balance its potential benefits against its potential risks in light of the latest scientific evidence.

Chapter Summary

- Salvia divinorum is a powerful psychoactive plant used by the indigenous people of Mexico for centuries.
- The active ingredient in salvia is salvinorin A, a potent and selective kappa opioid receptor agonist.
- The effects of salvia can vary greatly, ranging from mild to intense, and can include alterations in perception, thought, mood, and consciousness.
- Salvia has potential medical benefits, including treating addiction, depression, and pain.
- Salvia also has the potential to be abused, and individuals should approach its use with caution.

- Common side effects of salvia include nausea, dizziness, headache, loss of coordination, and impaired judgment.
- The effects of salvia can be unpredictable and dangerous, especially for individuals with a history of mental health problems.
- The use of salvia should be guided by caution and the latest scientific evidence.

8

DISSOCIATIVE DRUGS AND MENTAL HEALTH

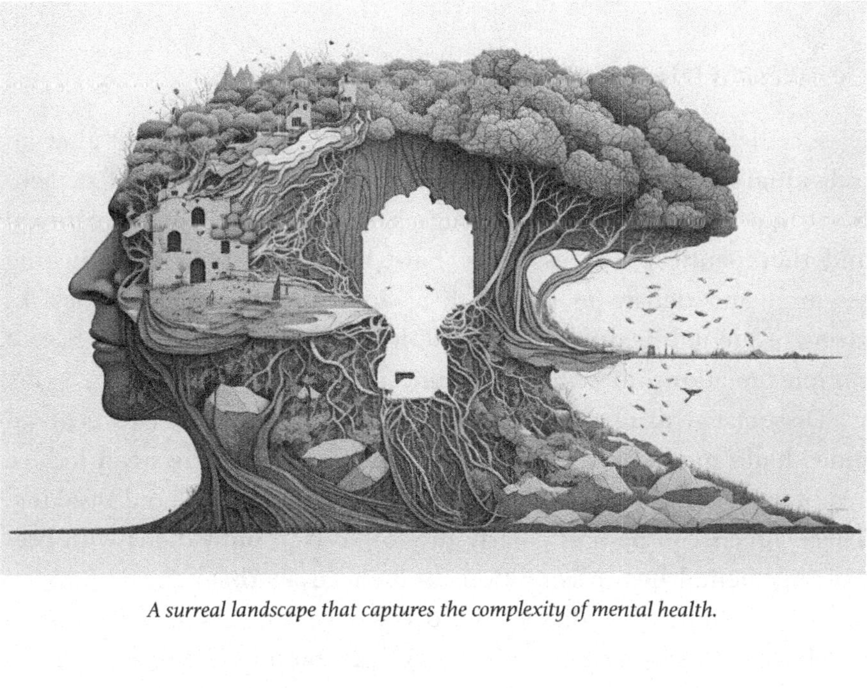

A surreal landscape that captures the complexity of mental health.

Dissociative drugs have long been a topic of interest in scientific and popular communities. These drugs can alter an individual's perception of reality and have been used for recreational and therapeutic purposes. The relationship between dissociative drugs and mental health is complex and multifaceted. This chapter will explore the relationship between dissociative drugs and mental illness, the symptoms these drugs may induce or exacerbate, and their potential role in treating mental illness. However, it's also important to examine the potential dangers of using dissociative drugs as a treatment, as they can have serious and long-lasting effects on one's mental and physical well-being. Through thoroughly examining the latest research and expert opinions, this chapter aims to provide a comprehensive overview of the relationship between dissociative drugs and mental health.

Overview of Dissociative Drugs and Mental Health

Dissociative drugs are a class of psychoactive substances that alter an individual's perception of reality and can induce feelings of detachment and dissociation. These drugs have been used for recreational and therapeutic purposes. They have been the subject of ongoing research and debate in the medical and scientific communities. In terms of mental health, dissociative drugs have been found to have a complex relationship with mental illness.

Dissociative drugs can have short-term and long-term effects on an individual's mental health. In the short term, these drugs can induce symptoms such as hallucinations, delusions, and disordered thinking. These effects can be particularly pronounced in individuals with pre-existing mental health conditions and can exacerbate existing symptoms. In the long term, dissociative drugs have been associated with the development of persistent symptoms of mental illness, such as depression, anxiety, and paranoia.

It's also important to note that dissociative drugs can have physical health consequences, such as heart palpitations, muscle spasms, and tremors. Moreover, they can have serious and long-lasting effects on an individual's brain function, including memory impairment and cognitive difficulties.

The relationship between dissociative drugs and mental health is complex and multifaceted. Therefore, it's essential to continue researching this topic and understanding the potential consequences of these drugs, both positive and negative, to provide informed and effective treatment for individuals with mental illness.

The Relationship Between Dissociative Drugs and Mental Illness

The relationship between dissociative drugs and mental illness is a complex and controversial issue that has been the subject of ongoing research and debate in the medical and scientific communities. On the one hand, some individuals with mental illness may use dissociative drugs as a form of self-medication to relieve symptoms of their condition. On the other hand, dissociative drugs can have negative consequences. It can exacerbate existing symptoms or even lead to new mental health problems.

Studies have shown that individuals with certain mental illnesses, such as depression, anxiety, and post-traumatic stress disorder (PTSD), may be more likely to use dissociative drugs as a form of self-medication. This is because dissociative drugs can induce feelings of detachment and dissociation, providing temporary relief from mental illness symptoms.

However, dissociative drugs can also lead to the development of persistent and persistent symptoms of mental illness, such as depression, anxiety, and paranoia. In some cases, long-term use of dissociative drugs can even lead to new mental health problems, such as psychotic disorders.

It's also important to note that individuals with pre-existing mental health conditions are at a higher risk of experiencing adverse effects from dissociative drugs. These effects can include increased symptoms of anxiety, depression, and paranoia, as well as an increased risk of self-harm and suicide.

In conclusion, the relationship between dissociative drugs and mental illness is complex and multi-faceted. While some individuals may use dissociative drugs as a form of self-medication, these drugs can have serious and long-lasting consequences on one's mental and physical well-being. Therefore, it's essential to continue researching this topic to provide informed and effective treatment for individuals with mental illness.

Dissociative Drugs and Symptoms of Mental Illness

Dissociative drugs have been found to have a complex relationship with mental illness, and their use can induce both short-term and long-term symptoms of mental illness. In the short term, dissociative drugs can induce symptoms such as hallucinations, delusions, and disordered thinking. These symptoms can be particularly pronounced in individuals with pre-existing mental health conditions and can exacerbate existing symptoms.

In the long term, dissociative drugs have been associated with the development of persistent symptoms of mental illness, such as depression, anxiety, and paranoia. In addition, studies have shown that individuals with pre-existing mental health conditions are at a higher risk of experiencing these symptoms after using dissociative drugs.

It's also important to note that dissociative drugs can lead to new mental health problems, such as psychotic disorders. In addition, in some cases, individuals may experience symptoms of mental illness even after they have stopped using dissociative drugs.

In addition to inducing symptoms of mental illness, dissociative drugs can also harm an individual's overall functioning. This can include memory and cognitive functioning difficulties, decreased motivation, and decreased ability to engage in daily activities.

In conclusion, dissociative drugs have been found to have a complex relationship with mental illness. They can induce both short-term and long-term symptoms of mental illness. In addition, using these drugs can also lead to new mental health problems and harm an individual's overall functioning. Therefore, it's essential to continue researching this topic to understand the relationship between dissociative drugs and mental illness and provide effective treatment for individuals affected by these conditions.

The Role of Dissociative Drugs in Treating Mental Illness

The role of dissociative drugs in treating mental illness is a controversial and complex issue that has been the subject of ongoing research and debate in the medical and scientific communities. While some studies have found that dissociative drugs may have therapeutic benefits for individuals with certain mental illnesses, such as post-traumatic stress disorder (PTSD) and depression, the potential dangers and negative consequences associated with the use of these drugs must be taken into consideration.

In the treatment of PTSD, for example, dissociative drugs have been found to have a calming effect on the central nervous system, reducing anxiety and fear. This can provide temporary relief from symptoms of PTSD, allowing individuals to engage in therapy and other forms of treatment.

However, it's important to note that the use of dissociative drugs in treating mental illness must be carefully monitored and used under the guidance of a medical professional. This is because the long-term effects of these drugs on an individual's mental and physical well-being are not yet fully understood, and individuals with pre-existing mental health conditions are at a higher risk of experiencing adverse effects.

In conclusion, the role of dissociative drugs in treating mental illness is a complex issue that requires ongoing research and careful consideration. While dissociative drugs may have therapeutic benefits for some individuals with certain mental illnesses, the potential dangers and negative consequences associated with the use of these drugs must be taken into account when making treatment decisions. Therefore, individuals with mental illness should only use dissociative drugs under the guidance of a medical professional and with a comprehensive and individualized treatment plan in place.

The Potential Dangers of Using Dissociative Drugs to Treat Mental Illness

The use of dissociative drugs to treat mental illness is associated with a range of potential dangers, including both short-term and long-term risks to an individual's mental and physical health.

In the short term, dissociative drugs can induce symptoms such as hallucinations, delusions, and disordered thinking. These symptoms can be particularly pronounced in individuals with pre-existing mental health conditions and can exacerbate existing symptoms. In addition, dissociative drugs can also result in impaired motor coordination and an increased risk of accidents, particularly in individuals who engage in activities such as driving or operating machinery.

Long-term use of dissociative drugs can lead to persistent symptoms of mental illness, such as depression, anxiety, and paranoia. Additionally, individuals who use dissociative drugs over an extended period are at a higher risk of developing a substance use disorder, such as addiction. This can lead to further negative consequences, including financial difficulties, social problems, and decreased overall functioning.

Furthermore, dissociative drugs can also hurt an individual's overall well-being, including difficulties with memory and cognitive functioning and decreased motivation and ability to engage in daily activities.

It's also important to note that the long-term effects of dissociative drugs on an individual's mental and physical health are not fully understood, and ongoing research is necessary to determine the full extent of these risks.

In conclusion, using dissociative drugs to treat mental illness is associated with various potential dangers, including short-term and long-term risks to an individual's mental and physical health. Therefore, individuals with mental illness who are considering dissociative drugs as a form of treatment should only use these drugs under the guidance of a medical professional and with a comprehensive and individualized treatment plan in place.

Chapter Summary

- Dissociative drugs alter an individual's perception of reality and have been used for recreational and therapeutic purposes.
- The relationship between dissociative drugs and mental health is complex and multifaceted.
- Dissociative drugs can have short-term and long-term effects on mental health, including inducing symptoms of mental illness and exacerbating existing symptoms.
- Dissociative drugs can also lead to new mental health problems and physical health consequences.
- The relationship between dissociative drugs and mental illness is controversial, with some individuals using them as a form of self-medication.
- Individuals with pre-existing mental health conditions are at a higher risk of experiencing adverse effects from dissociative drugs.
- Dissociative drugs can harm an individual's overall functioning and lead to memory and cognitive difficulties.
- Further research is necessary to understand the relationship between dissociative drugs and mental health to provide informed and effective treatment.

9

DISSOCIATIVE DRUGS AND SOCIETY

As the sun set, the cityscape was illuminated in a kaleidoscope of colors, reflecting the diversity and complexity of modern society.

Dissociative drugs have been a topic of **concern** for many years, affecting the individuals who use them and society. In this chapter, we will delve into the various impacts of dissociative drugs on crime and violence, the economy, healthcare, education, and more. We will also examine the government and law enforcement's role in addressing the use and abuse of dissociative drugs. With a clear and concise approach, this chapter aims to provide a comprehensive understanding of the complex issue of dissociative drugs and their impact on society.

Overview of Dissociative Drugs and Society

Dissociative drugs, also known as dissociatives, are a class of psychoactive substances that produce feelings of detachment and disconnection from one's surroundings and self. Some most commonly known dissociative drugs include ketamine, PCP, and DXM.

Dissociative drugs have been a topic of concern for society due to the potential harm they can cause to individuals and the wider community. These drugs can cause various negative effects, including hallucinations, confusion, agitation, and impaired motor function. High doses can also lead to serious health issues such as respiratory failure, coma, and even death.

Dissociative drugs have also been linked to increased crime and violence. In some cases, individuals under the influence of dissociatives have been involved in violent crimes, putting themselves and others at risk. The impact of dissociative drugs on the economy cannot be ignored either, as the cost of healthcare, law enforcement, and criminal justice resources used to address the problem is substantial.

In conclusion, dissociative drugs have significant and far-reaching effects on society. Therefore, it is important for individuals, communities, and the government to be aware of the dangers associated with these drugs and to work together to address the issue.

The Impact of Dissociative Drugs on Crime and Violence

The impact of dissociative drugs on crime and violence is a major concern for society. These drugs can induce confusion, agitation, and violence in users, leading to dangerous and unpredictable behavior. In some cases, individuals under the influence of dissociatives have committed violent crimes, putting themselves and others at risk.

Studies have shown that dissociative drugs are associated with an increased likelihood of criminal behavior, including assault, robbery, and domestic violence. For example, ketamine has been linked to violent behavior, and individuals who use PCP have been known to engage in unpredictable and sometimes violent acts.

The criminal justice system is also impacted by dissociative drugs, as law enforcement resources are often used to address incidents involving individuals under the influence of these substances. In addition, the cost of healthcare, rehabilitation, and legal proceedings associated with dissociative drug use can be substantial.

In conclusion, we cannot ignore the impact of dissociative drugs on crime and violence. Therefore, it is important for individuals, communities, and the government to be aware of the dangers of these drugs and take steps to prevent and reduce their use. This can include education, treatment, rehabilitation programs, and effective law enforcement measures to prevent the sale and distribution of these dangerous substances.

The Impact of Dissociative Drugs on the Economy

The impact of dissociative drugs on the economy is a significant concern for society. These drugs can lead to various negative consequences, including decreased productivity, increased healthcare costs, and decreased economic growth.

One of the key ways that dissociative drugs impact the economy is through decreased productivity. Individuals who use these drugs are more likely to miss work, struggle with concentration and motivation, and have **reduced job performance**. This can result in decreased productivity, lower earnings, and reduced economic growth.

Another major impact of dissociative drugs on the economy is increased healthcare costs. These drugs can lead to serious health problems, including respiratory failure, coma, and even death. This can significantly strain healthcare systems and increase government and healthcare providers' costs.

In addition, the cost of law enforcement and criminal justice resources used to address incidents involving individuals under the influence of dissociative drugs can also be substantial. This includes the cost of arrests, trials, and rehabilitation programs.

In conclusion, the impact of dissociative drugs on the economy is significant and far-reaching. Therefore, it is important for governments, communities, and individuals to be aware of these impacts and to take steps to prevent and reduce the use of these dangerous substances. This can include education, treatment, rehabilitation programs, and effective law enforcement measures to prevent the sale and distribution of dissociative drugs.

The Impact of Dissociative Drugs on Healthcare

The impact of dissociative drugs on healthcare is a major concern for society. These drugs can cause serious health problems, including respiratory failure, coma, and even death, putting a strain on healthcare systems and increasing costs for governments and healthcare providers.

One of the key ways that dissociative drugs impact healthcare is through their effects on the central nervous system. These drugs can cause hallucinations, confusion, and impaired motor function, leading to an increased risk of accidents and injuries. These accidents can sometimes result in serious health problems, including head injuries, broken bones, and organ damage.

Another way that dissociative drugs impact healthcare is through their potential to cause long-term health problems. These drugs can result in chronic health conditions, including memory problems, mood disorders, and psychological trauma. This can result in increased costs for healthcare providers and decreased quality of life for individuals who use these drugs.

In addition, the cost of treatment and rehabilitation programs for individuals affected by dissociative drugs can also be substantial. This includes the cost of hospitalization, therapy, and rehabilitation services.

In conclusion, the impact of dissociative drugs on healthcare is significant and far-reaching. Therefore, it is important for governments, communities, and individuals to be aware of these impacts and to take steps to prevent and reduce the use of these dangerous substances. This can include education, treatment, rehabilitation programs, and effective law enforcement measures to prevent the sale and distribution of dissociative drugs.

The Impact of Dissociative Drugs on Education

The impact of dissociative drugs on education is a major concern for society. These drugs can seriously affect individuals' mental and physical health, leading to decreased educational attainment and reduced long-term economic opportunities.

One of the key ways that dissociative drugs impact education is through their effects on cognitive function. These drugs can cause memory problems, confusion, and reduced motivation, making it difficult for individuals to concentrate and perform well in school. This can result in lower grades, decreased educational attainment, and reduced opportunities for future employment and financial stability.

Another way that dissociative drugs impact education is through their potential to cause long-term health problems. These drugs can result in chronic health conditions, including memory problems, mood disorders, and psychological trauma. This can make it difficult for individuals to maintain focus and perform well in school, leading to decreased educational attainment and reduced opportunities for future success.

In addition, the cost of treatment and rehabilitation programs for individuals affected by dissociative drugs can also be substantial. This includes the cost of therapy, rehabilitation services, and missed school days.

In conclusion, the impact of dissociative drugs on education is significant and far-reaching. Therefore, it is important for governments, communities, and individuals to be aware of these impacts and to take steps to prevent and reduce the use of these dangerous substances. This can include education, treatment, rehabilitation programs, and effective law enforcement measures to prevent the sale and distribution of dissociative drugs.

The Role of Government and Law Enforcement in Addressing the Use of Dissociative Drugs

The role of government and law enforcement in addressing the use of dissociative drugs is critical to reducing the negative impacts of these substances on society. These entities play a vital role in preventing the sale and distribution of dissociative drugs and providing education, treatment, and rehabilitation resources to individuals affected by these substances.

One of the key ways that governments and law enforcement can address the use of dissociative drugs is through effective law enforcement measures. This includes cracking down on illegal drug labs and distributors and intercepting and confiscating illegal drugs before they reach users. These efforts can help to reduce the supply of these drugs and make them less accessible to individuals who might use them.

Another way that governments and law enforcement can address the use of dissociative drugs is through the provision of education, treatment, and rehabilitation resources. This includes developing educational programs to raise awareness about the dangers of these drugs, as well as creating treatment and rehabilitation programs for individuals affected by these substances. These resources can help individuals overcome dissociative drugs' negative effects and regain their health and productivity.

In conclusion, the role of government and law enforcement in addressing the use of dissociative drugs is critical to reducing the negative impacts of these substances on society. In addition, effective measures to prevent the sale and distribution of these drugs, as well as education, treatment, and rehabilitation resources, are essential to reducing the harm caused by dissociative drugs and ensuring a safe and healthy society.

Chapter Summary

- Dissociative drugs are psychoactive substances that produce feelings of detachment and disconnection.
- Dissociative drugs have been linked to increased crime and violence and significantly affect the economy and healthcare.
- The impact of dissociative drugs on crime and violence includes violent and dangerous behavior and the use of law enforcement resources.
- Dissociative drugs impact the economy by decreasing productivity, increasing healthcare costs, and straining the criminal justice system.
- Dissociative drugs impact healthcare by causing serious health problems, increasing the risk of accidents and injuries, leading to chronic health conditions and increased treatment costs.
- It is important for individuals, communities, and the government to be aware of the dangers of dissociative drugs and to take steps to prevent and reduce their use.
- These steps include education, treatment, rehabilitation programs, and effective law enforcement measures.
- The impact of dissociative drugs on society is significant and far-reaching. It requires a comprehensive approach to address the issue.

10

DISSOCIATIVE DRUGS AND ADDICTION

A woman stands in the dark, completely enthralled by the small light of her phone. She can feel the pull of the addiction, yet she fights against it, knowing it can bring her nothing but darkness.

Dissociative drugs, also known as dissociatives, are a class of psychoactive substances that alter perception and consciousness. They have been used for centuries for their ability to create an altered state of mind. However, they also carry a high risk of addiction and negative side effects. Therefore, understanding the nature of dissociative drug addiction is crucial to help those affected by this condition. In this chapter, we will explore the physical and psychological effects of dissociative drug addiction, the development and progression of the disorder, the role of social and environmental factors, and the available treatment options. By delving into the science behind dissociative drug addiction, we aim to increase awareness, understanding, and compassion for those struggling with this complex issue.

Overview of Dissociative Drug Addiction

Dissociative drug addiction is a complex condition that involves the chronic use of dissociative drugs such as ketamine, PCP, and DXM. These drugs produce feelings of detachment from one's self, surroundings, and reality. In addition, they are often used for their mind-altering effects. Dissociative drugs block certain neurotransmitters in the brain, leading to changes in perception, mood, and consciousness.

However, the repeated use of dissociative drugs can lead to physical and psychological dependence and tolerance, where higher doses are needed to achieve the desired effects. This can result in a cycle of substance abuse, leading to addiction and damaging consequences on one's personal, professional, and social life.

Dissociative drug addiction can have severe health implications, including memory loss, anxiety, depression, and physical problems such as bladder and kidney damage. It can also increase the risk of developing mental health disorders and substance use disorders, making it a complex and challenging condition to treat.

In conclusion, dissociative drug addiction is a serious issue that can have severe and long-lasting consequences. Therefore, it is crucial to understand this condition's nature and be aware of the dangers of chronic dissociative drug use. The first step towards recovery is acknowledging that one has a problem and seeking help.

The Physical and Psychological Effects of Dissociative Drug Addiction

Dissociative drug addiction's physical and psychological effects can be severe and long-lasting. Dissociative drugs alter the balance of neurotransmitters in the brain, leading to changes in perception, mood, and consciousness. This can result in both short-term and long-term effects.

Short-term effects of dissociative drug use can include hallucinations, changes in perception of time and space, increased heart rate, increased blood pressure, and impaired motor function. In addition, dissociative drugs can sometimes cause dangerous side effects, such as overdose and potentially fatal respiratory depression.

Long-term effects of dissociative drug addiction can include persistent changes in perception, memory loss, depression, anxiety, and other mental health problems. Chronic dissociative drug use can also result in physical problems, such as bladder and kidney damage and cardiovascular problems.

In addition to the physical effects, dissociative drug addiction can also have severe psychological effects. Substance abuse can lead to a decline in one's personal and professional life, causing financial and relationship problems and feelings of hopelessness and helplessness. The cycle of substance abuse and addiction can be difficult to break, leading to a decline in physical and mental health and overall quality of life.

In conclusion, dissociative drug addiction's physical and psychological effects can be severe and long-lasting. Therefore, it is crucial to seek help at the earliest signs of substance abuse to prevent the negative consequences of addiction. In addition, treatment options such as behavioral therapy, medication, and support groups can help individuals overcome dissociative drug addiction and achieve a healthier, more fulfilling life.

The Development and Progression of Dissociative Drug Addiction

Various individual and environmental factors can influence the development and progression of dissociative drug addiction. Substance abuse often starts as a form of recreational drug use. Still, it can quickly escalate into a dangerous cycle of addiction.

One of the key factors in the development of dissociative drug addiction is tolerance. Over time, the brain becomes accustomed to the presence of the drug, leading to a decrease in its effectiveness. This can increase drug use as individuals seek to achieve the same effects they did when they first started using. This increase in drug use can lead to physical and psychological dependence and, eventually, addiction.

Social and environmental factors can also play a role in the development of dissociative drug addiction. For example, substance abuse can often stem from a desire to escape stressful life circumstances, such as trauma, mental health problems, or relationship problems. Substance abuse can also be influenced by peer pressure and access to drugs.

The progression of dissociative drug addiction can be difficult to predict but is often characterized by a decline in physical and mental health and personal and professional life. In addition, substance abuse can lead to financial problems, relationship problems, and a decline in overall quality of life.

In conclusion, the development and progression of dissociative drug addiction are complex and influenced by a range of individual and environmental factors. Understanding these factors can help individuals recognize the warning signs of addiction and seek help before it is too late. In addition, early intervention can help prevent the negative consequences of addiction and increase the chances of a successful recovery.

The Role of Social and Environmental Factors in Dissociative Drug Addiction

Social and environmental factors play a significant role in the development and progression of dissociative drug addiction. Substance abuse can stem from a desire to escape life stressors, such as trauma, mental health problems, or relationship problems. Peer pressure, drug access, and cultural attitudes toward substance abuse can also contribute to the development of addiction.

Peer pressure can be a major factor in the initiation of drug use. It can influence an individual's decision to continue using drugs. Being surrounded by individuals who use drugs or having friends who are also addicted can increase the likelihood of substance abuse and addiction.

Access to drugs is also a major factor in the development of dissociative drug addiction. The availability of drugs and the methods of obtaining them can greatly influence an individual's decision to use and continue using drugs.

Cultural attitudes towards substance abuse can also play a role in the development of addiction. For example, substance abuse may be normalized or encouraged in some cultures, leading individuals to believe that drug use is acceptable or desirable.

In addition to these factors, environmental stressors can also contribute to the development of dissociative drug addiction. For example, individuals who experience trauma, poverty, or other forms of adversity may be more likely to turn to substance abuse as a means of coping.

In conclusion, social and environmental factors play a significant role in the development and progression of dissociative drug addiction. Understanding these factors is crucial to preventing substance abuse and supporting individuals in their recovery journey. Addressing these factors can also help reduce the stigma associated with substance abuse and improve access to effective treatment and support.

The Treatment Options for Dissociative Drug Addiction

The treatment of dissociative drug addiction can be complex and may involve a combination of different approaches. The most effective treatments for dissociative drug addiction typically involve a combination of behavioral therapy, medication, and support from family and friends.

Behavioral therapy is a critical component of dissociative drug addiction treatment. This therapy can help individuals understand and address the underlying factors contributing to substance abuse. Common behavioral therapies used in treating drug addiction include cognitive-behavioral therapy, contingency management, and motivational interviewing.

Medication can also be an effective treatment option for dissociative drug addiction. Some medications can help alleviate withdrawal symptoms, reduce cravings, and support recovery. For example, naltrexone and acamprosate effectively treat addiction to dissociative drugs.

Inpatient and outpatient treatment programs can also provide a supportive environment for individuals in recovery. These programs can offer a range of services, including individual and group therapy, medication management, and support from peers in recovery.

The support of family and friends is also an important factor in the recovery process. Encouragement and support from loved ones can help individuals stay motivated and on track in their journey toward recovery.

In conclusion, the treatment of dissociative drug addiction can be complex. It may involve behavioral therapy, medication, and support from loved ones. Seeking help as early as possible can greatly increase the chances of a successful recovery and improve overall health and well-being. It is important to find a treatment program tailored to individual needs and to work closely with a healthcare provider to ensure a comprehensive and effective treatment plan.

Chapter Summary

- Dissociative drugs are psychoactive substances that alter perception and consciousness and carry a high risk of addiction.
- Dissociative drug addiction is a complex condition involving chronic use of drugs such as ketamine, PCP, and DXM.
- Dissociative drug addiction can have severe physical and psychological effects, including memory loss, anxiety, depression, bladder and kidney damage, and an increased risk of mental health disorders.
- Various individual and environmental factors, such as tolerance, social and environmental factors, and stress, can influence the development and progression of dissociative drug addiction.
- Short-term effects of dissociative drug use can include hallucinations, changes in perception, and impaired motor function. In contrast, long-term effects can include memory loss, depression, anxiety, and physical problems.
- The progression of dissociative drug addiction is often characterized by a decline in physical and mental health and personal and professional life.
- Social and environmental factors, such as peer pressure, drug access, and stress, play a significant role in the development and progression of dissociative drug addiction.
- Early intervention and treatment options, such as behavioral therapy, medication, and support groups, can help individuals overcome dissociative drug addiction.

AFTERWORD

In conclusion, this book has provided a comprehensive overview of dissociative drugs, including their history, chemical properties, medical uses and abuses, risks and side effects, laws and regulations, and personal experiences. It has also explored dissociative drugs' relationship with mental health, society, and addiction.

Dissociative drugs can have both positive and negative effects on individuals and society, and it is important to understand their impact. However, the risks associated with dissociative drugs can be serious, and it is crucial to take precautions to prevent misuse.

Ongoing research is being conducted better to understand dissociative drugs' effects and potential medical uses. In addition, new laws and regulations are being developed to address the misuse of these substances.

Finally, it is important to remember that individuals should only use dissociative drugs under the supervision of a medical professional. It is also important to educate yourself about the risks and dangers of these substances and seek help if you or someone you know is using them.

ACKNOWLEDGMENTS

I am grateful to many people who have supported me throughout the process of writing this book. First and foremost, I would like to express my gratitude to my family and friends for their love, encouragement, and patience during this journey. Without their unwavering support, this book would not have been possible.

I would like to extend my sincere thanks to the experts and researchers who have dedicated their lives to studying hallucinogens and dissociatives. Your insights and knowledge have been invaluable to the creation of this book.

I would also like to thank the publishers and editors who have helped me bring this project to fruition. Your expertise and guidance have been instrumental in shaping this book into its final form.

Finally, I want to thank the readers who have taken the time to delve into this topic. It is my hope that this book will provide you with a deeper understanding of hallucinogens and dissociatives and their potential impact on individuals and society.

ABOUT THE AUTHOR

Terence Wright is a researcher, writer, and expert on psychedelics, focusing on hallucinogens' effects and cultural significance. He has studied the subject for several years and has published several articles on the topic. Terence has a deep passion for understanding how psychedelics have been used throughout history and how they can be used in the future. He is an advocate of the responsible use of psychedelics. He believes that with more research and understanding, these substances can positively impact mental health and well-being. When not writing or researching, he enjoys hiking and spending time with his family.

FROM THE AUTHOR

Dear Reader,

I hope you enjoyed the book! I would love to hear your feedback—I personally read every review. If you have three minutes to help me, can you please post a short review on the platform from which you purchased this book?

Also, I would be honored if you would sign up for my mailing list to keep updated on my upcoming books, exclusive content, special promotions, and events. Plus, as a subscriber, you'll have the opportunity to **download my new books for free** in exchange for an honest review. Your feedback is important to me and will help me improve my books and reach a wider audience. Scan to QR code to sign up.

SCAN ME

Best, Terence Wright

Made in the USA
Coppell, TX
05 November 2026

11315633R00108